MW01077886

Demystifying Mixed-Signal Test Methods

DEMYSTIFYING MIXED-SIGNAL TEST METHODS

MARK BAKER

An imprint of Elsevier Science

Amsterdam Boston London New York Oxford Paris
San Diego San Francisco Singapore Sydney Tokyo

Newnes is an imprint of Elsevier Science.

Recognizing the importance of preserving what has been written, Elsevier Science prints its books on acid-free paper whenever possible.

Library of Congress Cataloging-in-Publication Data
Baker, Mark.
Demystifying mixed signal test methods / Mark Baker.
p. cm.
Includes index.
ISBN 0-7506-7616-7
1. Automatic checkout equipment. 2. Electronic apparatus and appliances—Testing 3. Mixed signal circuits. 4. Signal generators. I. Title.

TK7895.A8B34 2003
621.3815′48—dc21 2003041860

British Library Cataloguing-in-Publication Data
A catalogue record for this book is available from the British Library.

The publisher offers special discounts on bulk orders of this book.
For information, please contact:

Manager of Special Sales
Elsevier Science
200 Wheeler Road
Burlington, MA 01803
Tel: 781-313-4700
Fax: 781-313-4882

For information on all Newnes publications available, contact our World Wide Web home page at: http://www.newnespress.com

10 9 8 7 6 5 4 3 2 1

Transferred to Digital Printing 2006

To Richone

I love being married to you!

CONTENTS

PREFACE

Technology is a useful tool; but that's all it is. It seems to me that we can get caught up in our own cleverness to the point where we neglect the value of the human soul. I hope you find the following paraphrase from the Gospel of Luke to be a gentle encouragement to treasure what is truly precious.

"Do not worry about your life, what you will eat; nor about the body; what you will put on. Life is more than food, and the body is more than clothing. Consider the ravens, for they neither sow nor reap, nor do they have cell phones. They have no bank accounts or portfolios, and God feeds them. Of how much more value are you than the birds?"

"And which of you by worrying can add one penny to his net worth? If you then are not able to do the least, why are you anxious for the rest? Consider the lilies of the field, for they neither execute a business plan nor go online, yet even the venture capitalists in all their glory are not arrayed as one of these. If then God so clothes the grass, which today is in the field and tomorrow is thrown into the oven, how much more will He care for you? And do not seek to get more and more stuff, nor have an anxious mind; for all of these things Wall Street seeks after; and your Father knows what things you need. But seek the Kingdom of God, and all these things shall be added to you."

CHAPTER 1

INSIDER'S GUIDE TO MIXED SIGNAL TEST

This, gentlemen, is a football.

—Vince Lombardi

Hi. My name's Mark Baker, and I've been working in and around mixed signal test for longer than I'd like to remember. After I'd been presenting seminars about mixed signal test for a few years, people said I should write a book. So here it is, and I hope you like it.

A while back, I caught up with my brother Len at some airport, somewhere. Len's a banker, and when I gave him my business card, he looked it over and laughed, "Mixed Signal? What the heck is mixed signal? It sounds like the messages I get from my ex-wife!" Fortunately, the kind of mixed signal we're talking about here has to do with electronics, a heck of a lot simpler subject than that other kind of mixed signal.

1.1 What Is Mixed Signal?

My favorite illustration of mixed signal technology is a music CD, which used to be known as a "Compact Digital disc." The optically encoded information is digital data—ones and zeroes—but there is not an op-code or an address fetch to be found. Even more remarkable, when you put this Compact Digital disc into a CD player, music comes out. In this example of mixed signal, analog information is processed in digital form.

A modem is another common technology that illustrates mixed signal technology. If you've ever listened in on those awful squeaks and squawks that come out of a modem, you know there's something strange going on. A modem interfaces between an analog system (the phone line) and a digital system (the data you are transmitting from your computer). In the case of a modem, digital information is processed in analog form.

These two simple illustrations of a music CD and a modem lead to a general definition of mixed signal technology: A mixed signal system (or component)

- processes analog information in digital form; or
- processes digital information in analog form; or
- both.

Mixed signal technology forges analog and digital together into a powerful combination with tremendous potential. The ubiquitous cell phones and CD players are everyday examples of the importance and significance of mixed signal. But, how do we go about testing mixed signal devices? I'll let you in on a secret, if you promise not to tell. Mixed signal test is not all that hard. It's different, sure, but it's not like you have to be some kind of rocket scientist to make this stuff work. Let's begin by dispelling some myths:

Myth: Mixed signal test engineers are born, not made.
Fact: Mixed signal test is not any more difficult than logic test or analog test. It's just different.

Myth: Mixed Signal test requires graduate-level math skills.
Fact: Computers are good at doing complex math. Slide rules and pocket protectors are not required.

Myth: You have to be some kind of guru.
Fact: Do I sound like a guru to you?

Myth: Mixed signal test parameters are weird and fuzzy.
Fact: Mixed signal test parameters often describe *signal characteristics* as parametric values. It's not unlike testing supply current on a logic device.

Myth: You have to wear a long robe, a pointed hat, and carry a magic wand.
Fact: The wand is not necessary. Most mixed signal test engineers look and sound like normal people, until they start talking about FFTs. (As with all test engineers, a copy of *Ye Olde Book of Spelles and Incantations* is standard equipment.)

It's like anything else. Master the fundamentals, practice, and eventually you'll get to be pretty good. Mastering the fundamentals is what this book is about. Just don't panic or go into brain-lock. Remember that movie, *What About Bob?* There was a pearl of wisdom beneath all that hilarity, which is "baby steps." Take it slow, have some fun, and you'll do just fine.

1.2 Some Terminology

What do we mean by "digital" and "analog"? Well, the word "digital" comes from the word "digits," which in turn is Latin for "fingers." No kidding! Using the "digits" on our hands was the first way we learned how to count, add, and subtract. The term "digital," therefore, has to do with processing information that is in numerical form, or discrete units. Addition and subtraction of whole numbers is a digital process.

The term "analog" has the same root as "analogy." Analog uses a representation or equivalent and deals with information as a continuum. A microphone that converts sound into electrical signals is analog.

Before we close the dictionary, let's look at the term "signal." (If we're working with mixed signal, wouldn't it be a good idea to know what a signal is?)

> A detectable (or measurable) physical quality or impulse (as voltage, current, or magnetic field strength) by which messages or information can be transmitted. (from Webster's New Collegiate Dictionary.)

1.2.1 What's an Analog Signal?

Analog signals concern electrical variables, their rate of change, and the associated energy levels. Any point on an analog signal may be at any level within a given range.

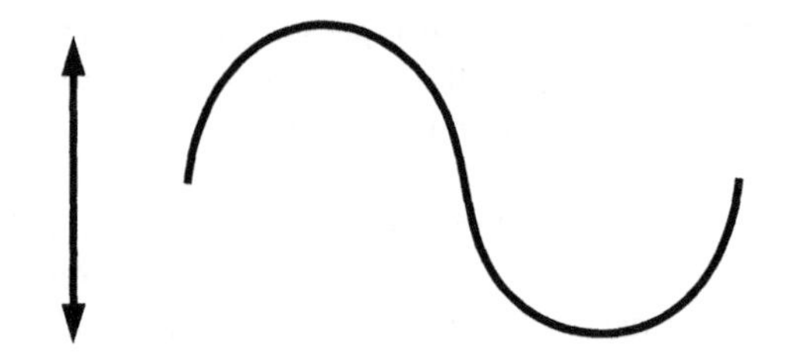

Figure 1.1

Analog Information—A Continuum

1.2.2 What's a Digital Signal?

In this discussion, digital signals concern electrical variables that have been formatted to represent binary digits. Any point on a digital signal will be either at one of two levels, representing a binary 0 or binary 1; at some other level representing no data; or at a transition from one level to another.

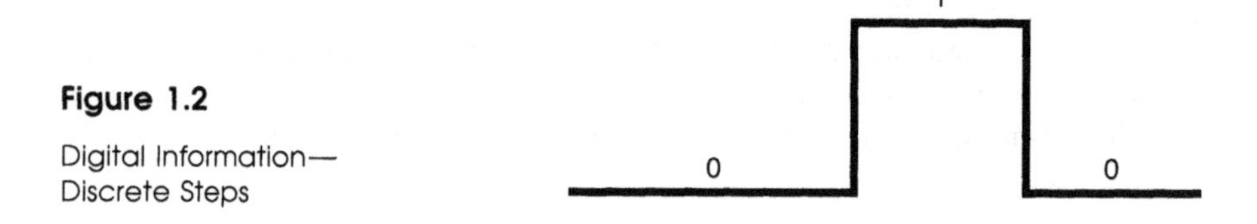

Figure 1.2

Digital Information—Discrete Steps

1.2.3 However...

Digital is not always logic. For example, on a music CD (compact disc), analog information is encoded in digital form. And, as you might guess, logic is not always digital, Analog circuits can add, subtract, multiply, divide, and perform log calculations. Your car stereo includes a dual analog multiplier—it's called an amplifier.

Sometimes, digital and analog are merged together, and that's what mixed signal is all about. As just noted, the digital data on a music CD represents *analog* information in *digital* form. Or, consider an Analog-to-Digital converter (ADC), which generates digital values corresponding to analog input levels. The output of an Analog-to-Digital converter is digital, but it's not logic. The output of an ADC is analog information that is represented in binary form.

There's also the opposite case, where we encounter digital information in analog form. The transmit side of a modem converts digital data (like the ASCII text of an email) into modulated analog signals. Higher frequency data transmission gets an analog spin as well. An Ethernet transceiver shapes digital data into bipolar analog pulses with controlled slopes. Going back to the component level, let's remember that the input to a Digital-to-Analog converter (DAC) is binary, but not logic. The input sequence to a DAC is simply analog information in digital form.

1.3 What Is a Mixed Signal Device?

Excellent question![1] Here's an attempt at a definition:

> A mixed signal device operates across digital and analog domains by representing or processing either analog or digital information in either analog or digital form.

Consider a matrix that identifies a device according to function type and data type. Mixed signal devices cross diagonal boundaries of a process/data matrix.

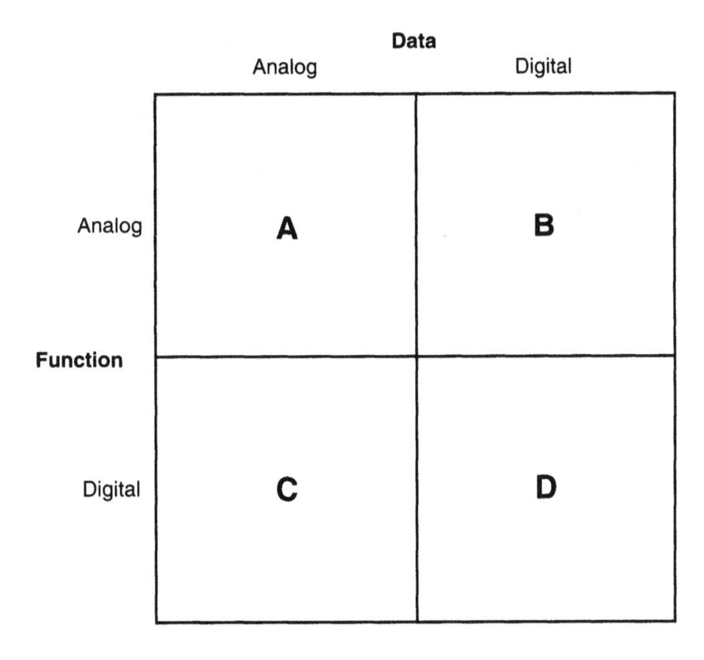

Figure 1.3

Function and Data Matrix

[1] There are two types of questions that I get from folks who attend my seminars—*good* questions, and *excellent* questions. Excellent questions are those to which I happen to know the answer.

We would probably agree that an op amp is a "pure analog" type of device An op amp processes analog data, and performs an analog function. So it goes in quadrant "A" in Fig. 1.3. A NAND gate processes digital data, and performs a digital function, which puts it in quadrant "D." A device that processes analog data with a digital function, or processes digital data with an analog function, is a mixed signal device. A mixed signal device crosses the boundaries between pure analog and pure digital.

Let's try a few examples to get the hang of it. Which quadrant, or category, would you assign to the following device types?

Category "A": Pure Analog
Category "B": Digital data and Analog function (analog data in digital form)
Category "C": Analog data and Digital function (digital data in analog form)
Category "D": Pure Digital

Table 1.1

Function and Data Product Types

Device Type	Category
Analog-to-Digital Converter	
Modem	
Digital-to-Analog Converter	
Music CD player	

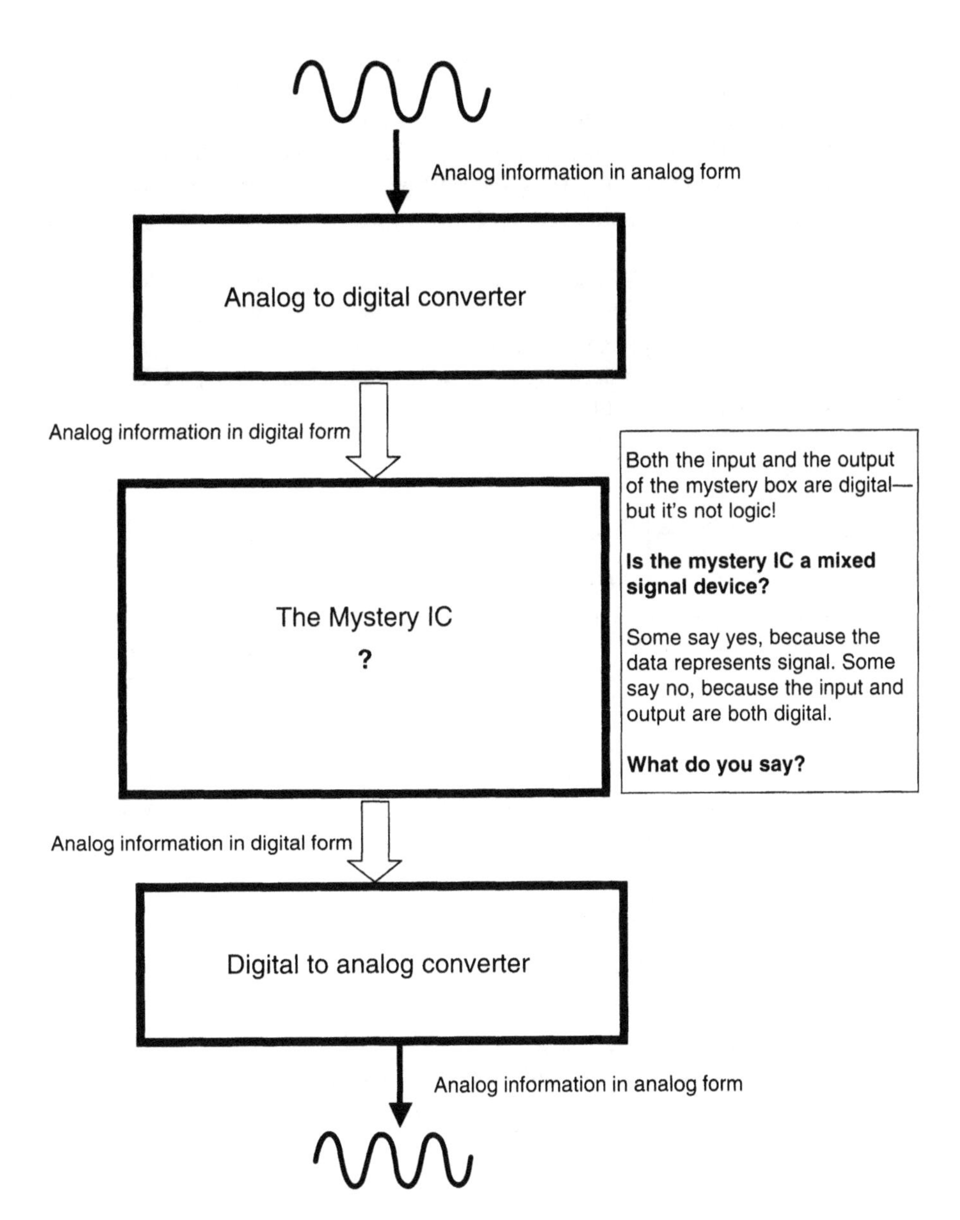

Figure 1.4

Crossing the Boundaries—DSP

1.4 What Isn't Mixed Signal?

If you take these definitions to an extreme, you can make an argument that *any* integrated circuit is a mixed signal device. Deep inside a microprocessor are millions of transistors that are behaving in a very analog manner. Generally, we'll see that mixed signal devices are loosely categorized according to the input/output pin function. If a device has analog and digital input and/or input pins, we'll call it mixed signal. On the other hand, what if a device receives analog information in digital form, and then processes the signal information using digital techniques? If the output of this device is analog information in digital form, then the i/o pins are all digital. But the function is analog. A Digital Signal Processor (DSP) is such a device.

I've never heard of a company that comes out with their new slower and simpler part, have you? Developing additional features sometimes includes a higher level of integration. A cell phone, for example, might initially contain three integrated circuits for the voice-band processing—an input amplifier and A/D, a DSP device, and an output DAC and amplifier.

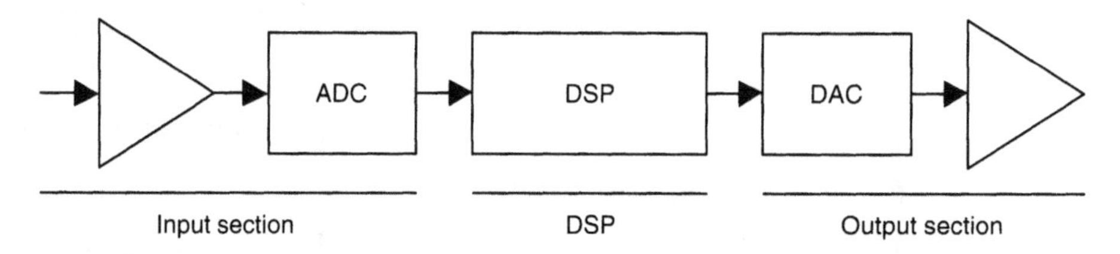

Figure 1.5

Analog In, Analog Out

If all three sections are integrated into a single device, what do you have? Analog in, analog out—and digital signal processing in the middle.

1.5 Example of Mixed Signal Devices

There are many mixed signal device categories, including

- Telecommunications
- Micro-Controllers with embedded analog
- Digital-to-Analog and Analog-to-Digital Converters
- Video Processing
- Interface
- DSP

1.5.1 Telecommunications Devices—Modems and Codecs

Telecommunications, or "telecom," devices are used to transmit and receive audio, digital, or video information. The telephone system is one example of a telecommunications network, designed for voice communication and using a device called a codec. In another application, a mixed signal device called a **modem** interfaces between a computer and the phone line in order to transmit digital data on the analog-based phone system.

Codecs

In order to efficiently process the analog signal information, the telephone system uses a device called a codec. Digital data is easier to transmit and store, and the purpose of the codec is to convert the audio voice analog signal into digital form. Codec is an abbreviation for "encoder and decoder." Analog signal information is applied to the transmit, or encoder, section, which generates a digital bit stream. The receive, or decoder, section receives the digital bit stream, and reconstructs the analog signal information. From a simplified viewpoint, a codec can be described as a combination of an analog-to-digital converter and a digital-to-analog converter.

If you make a phone call from California to Connecticut, the voice information is actually transmitted in digital form. Codecs at either end of the connection are the interface between the analog phone signal and the digital phone system.

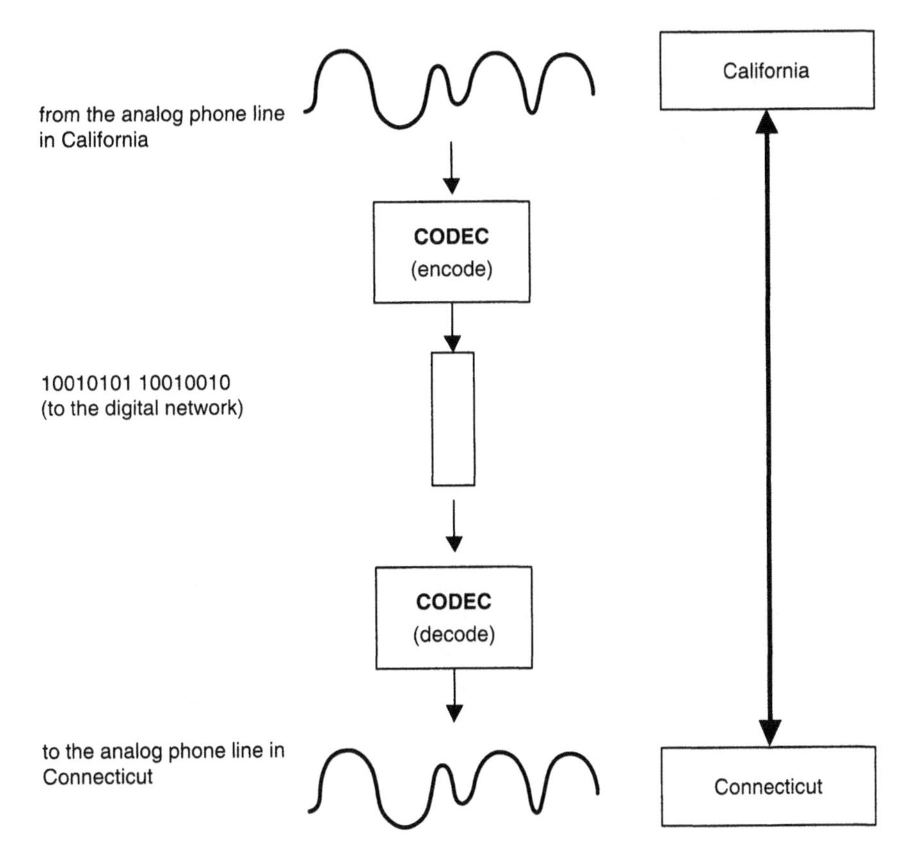

Figure 1.6

Codec Functionality: Analog Data in Digital Form

Modems

Modem is an abbreviation for "modulate and demodulate." The "modulate" section of a modem transmits digital data by varying the phase and amplitude of an analog signal. The "demodulate" section of a modem receives digital data by detecting the variations in the modulated analog signal.

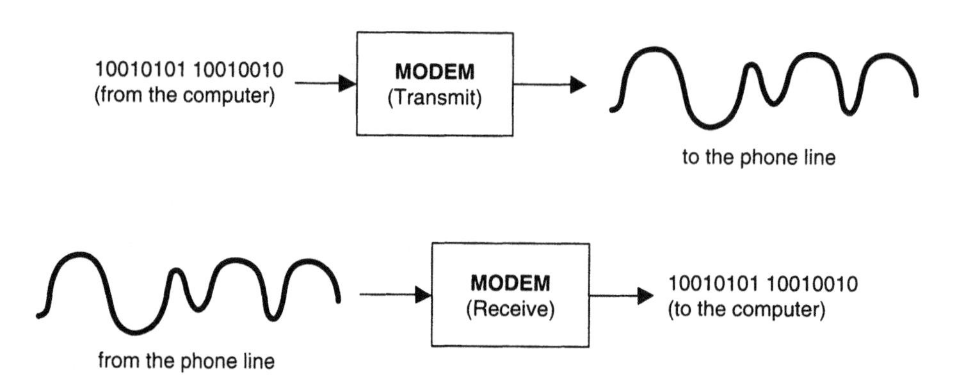

Figure 1.7

Modem Functionality: Digital Data in Analog Form

1.5.2 Micro-controllers with Embedded Analog

The ubiquitous micro-controller is found in applications ranging from industrial processes to consumer products such as microwave ovens and anti-lock brake systems. A powerful variation of the micro-controller chip includes on-chip analog-to-digital and digital-to-analog converters. This allows the controller to respond to analog input signals, and to produce a response in analog form. This combination of analog processing with digital logic illustrates a device category known as "digitally complex mixed signal."

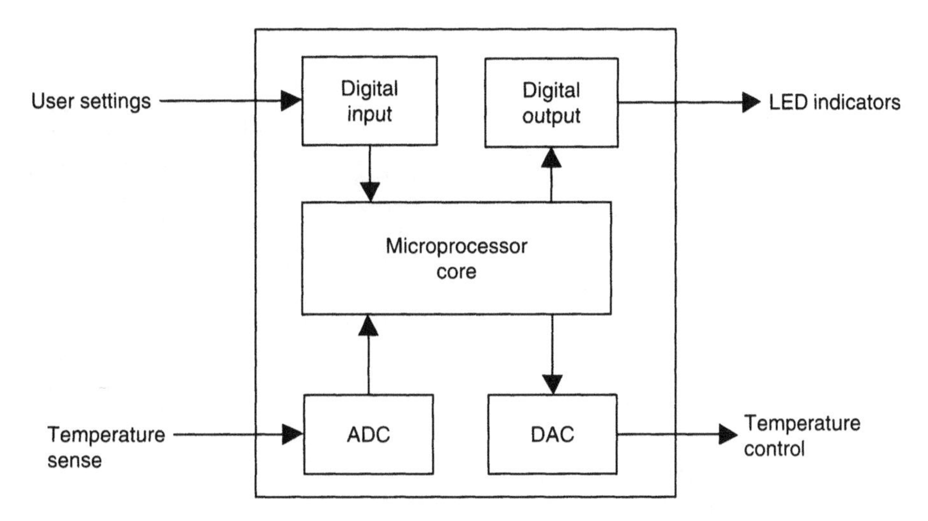

Figure 1.8

Microprocessor with Embedded Mixed Signal Functions

Suppose this is the micro-controller in an automatic bread-maker unit. The digital input port is used to read the various parameters that have been programmed by the user. The ADC input monitors the temperature of the bread dough. LED indicators are controlled by the digital output port, and the DAC output controls the bread-maker temperature.

During operation, the microprocessor core executes a program that reads the user parameters and the temperature data. In this example, the microprocessor program determines that, based on the user parameters, the temperature is too high. The program, in turn, sends a value to the digital output port that activates the "Over Temperature" LED and sends a digital value to the on-chip DAC to reduce the heat.

1.5.3 Discrete Converters

Converters are perhaps the most readily recognized type of mixed signal device. Analog-to-Digital converters accept an analog input, and generate a corresponding digital output code. Digital-to-Analog converters accept a digital input and generate an analog output level.

The design constraints for converters creates an inverse relationship between the analog precision and the conversion rate. Very high precision converters may feature an analog precision of one part per million, but have relatively slow conversion rates. High-speed converters may operate at several hundred megahertz, but tend to have relatively low analog precision. Depending on the end-use application, there are trade-offs concerning the required precision, speed, and cost. High-accuracy converters are found in audio and industrial control applications, whereas instrumentation and video applications utilize high-speed converters.

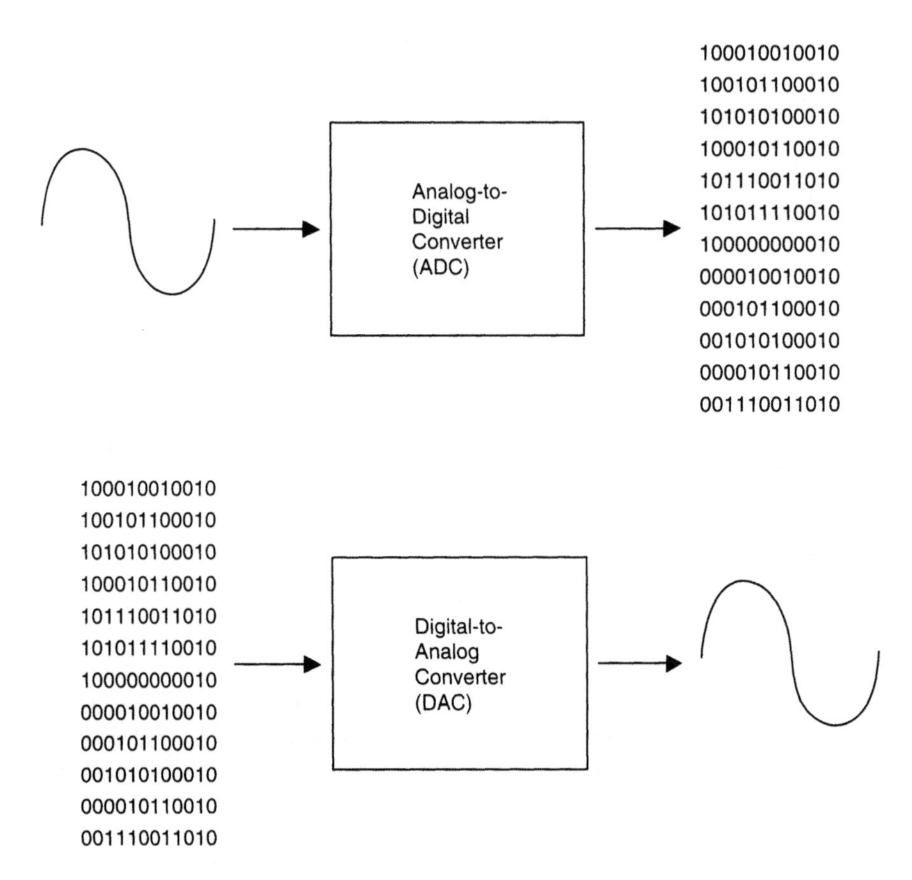

Figure 1.9

Data Converters

1.5.4 Mass Storage Interface

The hard disk drive unit of a personal computer requires extremely sophisticated mixed signal processing. The rotating media of the hard disk contains magnetically encoded digital data. The magnetic charges, arranged in concentric tracks of the disk, are sensed by positioning a coil mechanism, called the read head, over the track. As the disk rotates, the magnetic charge on the track creates an inductive pulse across the coil mechanism. The inductive pulse, in turn, generates current flow across the coil. The current pulse across a resistor produces a voltage signal.

The voltage pulse train from the disk read head is of low amplitude and high speed, and also contains a large amount of extraneous information, or noise. Variations in the location of the track cause corresponding variations in the amplitude and frequency of the digitally encoded analog signal. The analog signal frequency may be several hundred megahertz, with average peak amplitudes of less than 1 mV. Some of the mixed signal functions employed to extract the digital data from a hard disk include amplification, noise reduction, pulse detection, synchronization, and decoding.

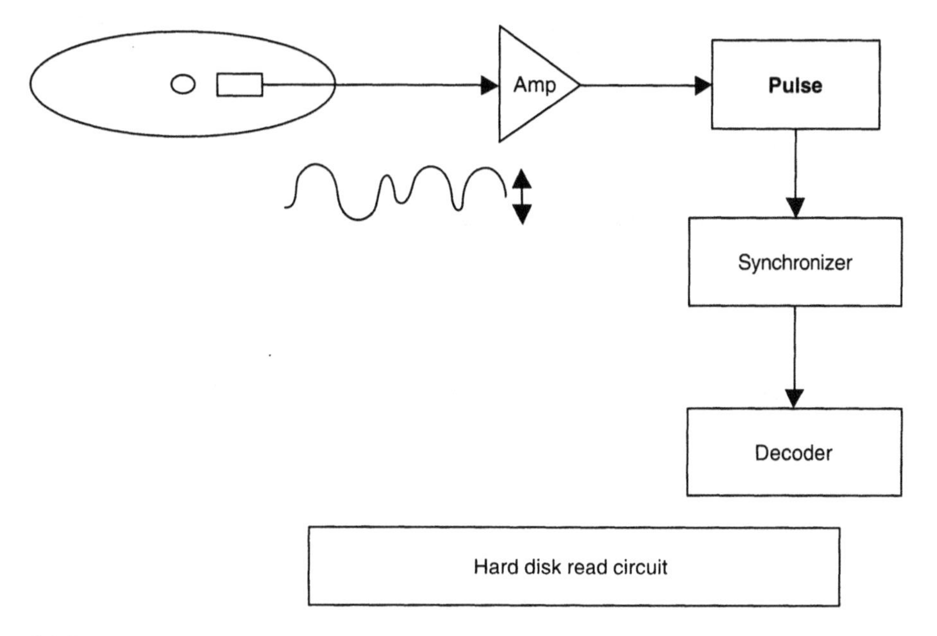

Figure 1.10

Hard Disk Read Circuit

1.5.5 Ethernet Transceivers

Ethernet transceivers send and receive digital data in analog form. The constraints of the cable interconnect require that the transmitted digital bit stream be formatted to minimize the capacitance and transmission line effects. The receive side of the Ethernet transceiver must be able to properly extract the digital data from the degraded signal.

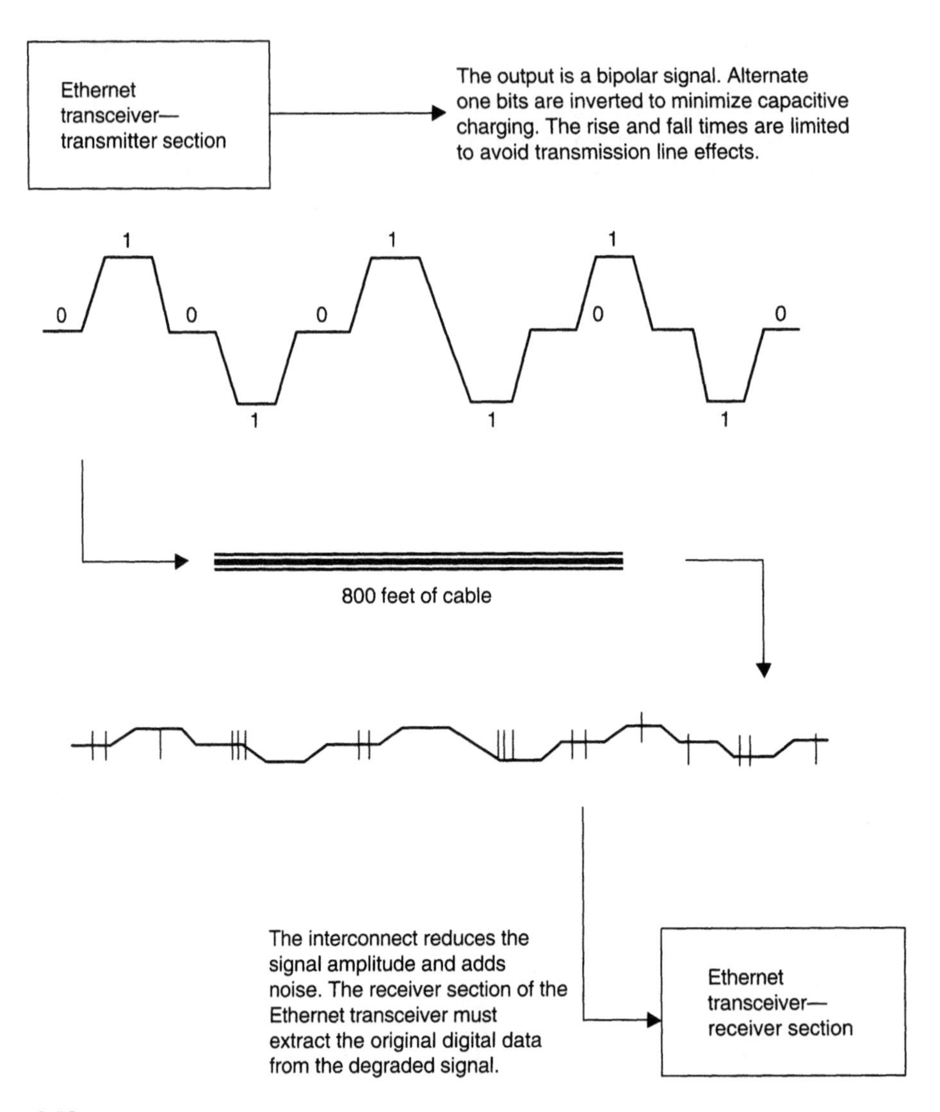

Figure 1.11

Ethernet Transceiver

1.6 Automatic Test Equipment (ATE)

Now that we've got a general sense of what mixed signal devices are like, we can start to look at what mixed signal test is all about. To understand the architecture of a mixed signal test system, let's look at automatic test equipment in general, and then examine logic test and analog test systems. Mixed signal test equipment includes characteristics of both digital and analog test systems.

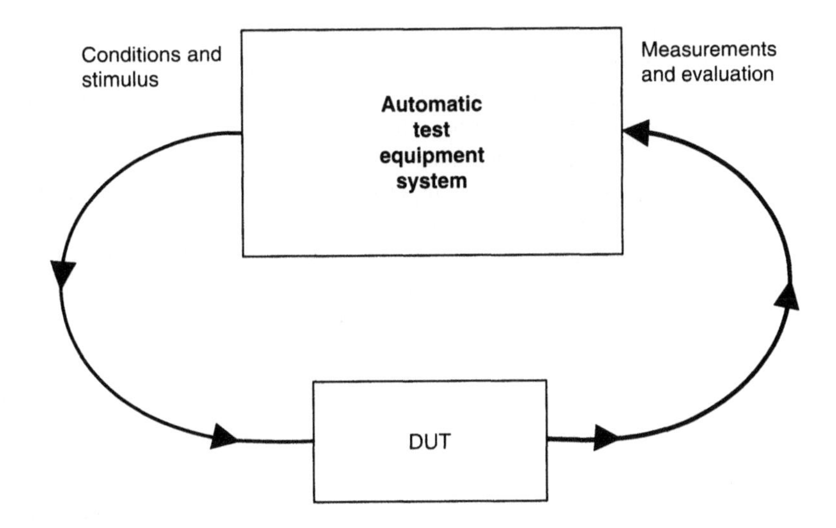

Figure 1.12

Fundamental Purpose of Automatic Test Equipment (ATE)

The **conditions and stimulus** provided by the ATE system include

power supply and ground
input signals

The **measurements** performed by the ATE system include

input pin impedance
input logic level thresholds
output pin voltage and compliance
rise and fall time
propagation delay
functional speed
output signal

The **evaluation** performed by the ATE system includes

Adequate DC performance?
Adequate AC performance?
Proper Logic function?
Adequate functional speed?
Proper signal characteristics?

1.6.1 ATE System Components

The three primary components of an ATE system are the mainframe, the test head, and the workstation. The mainframe is the largest section, and contains the power supplies, DC system, logic system, and measurement instruments. The tester resources that are contained in the mainframe are connected to the test head via a cable set. The test head is designed to be small and light enough to easily interface with a handler or prober. Critical test hardware components that must be physically close to the Device Under Test (DUT), like the driver/receiver circuits, are located in the test head. The test head features a connector section that makes all of the test system resources available to the DUT. A custom circuit board, called the load board or device interface board, completes the connection path from the test head connection points to the DUT pins. The workstation is the computer system that runs the test program and serves as the interface between the test system and the user. Some test systems use a second computer, located in the mainframe, as the test system controller, while the workstation is dedicated to the user interface.

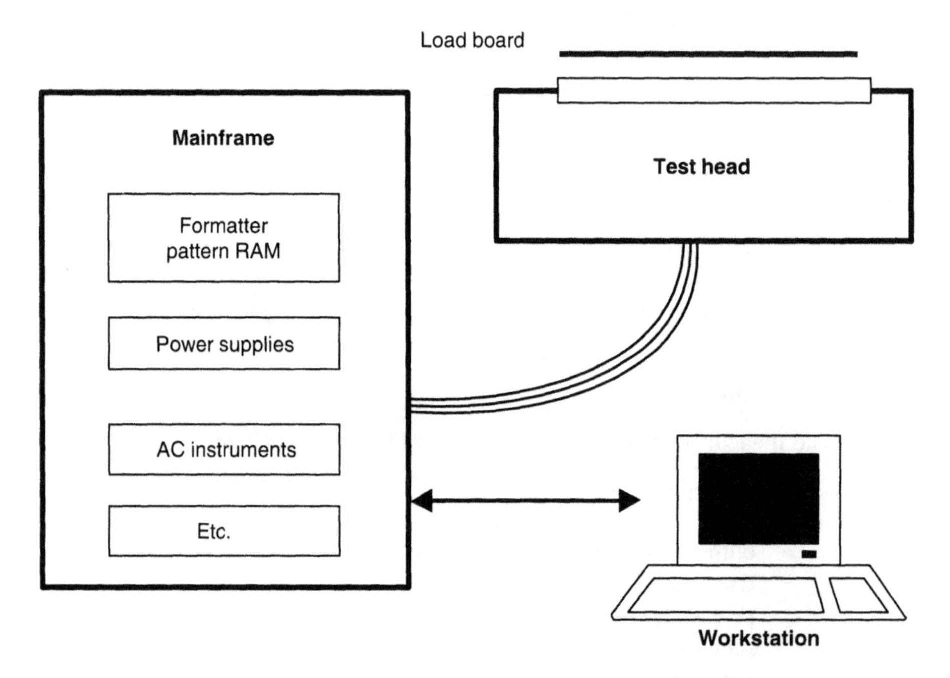

Figure 1.13

ATE System Components

1.7 Comparing Logic Test and Mixed Signal Test

Let's take a look at the different types of test systems, starting with logic testers. A logic tester will *source* digital data, *capture* digital data, and *compare* digital data against a model of the device function (pattern). When testing digital Integrated Circuits (ICs), the conditions and measurements are based on a pattern, which is a representation of the device functionality. The test system applies the pattern sequence of logic levels to the device under test and measures the response of the device by comparing the output against the expected pattern data.

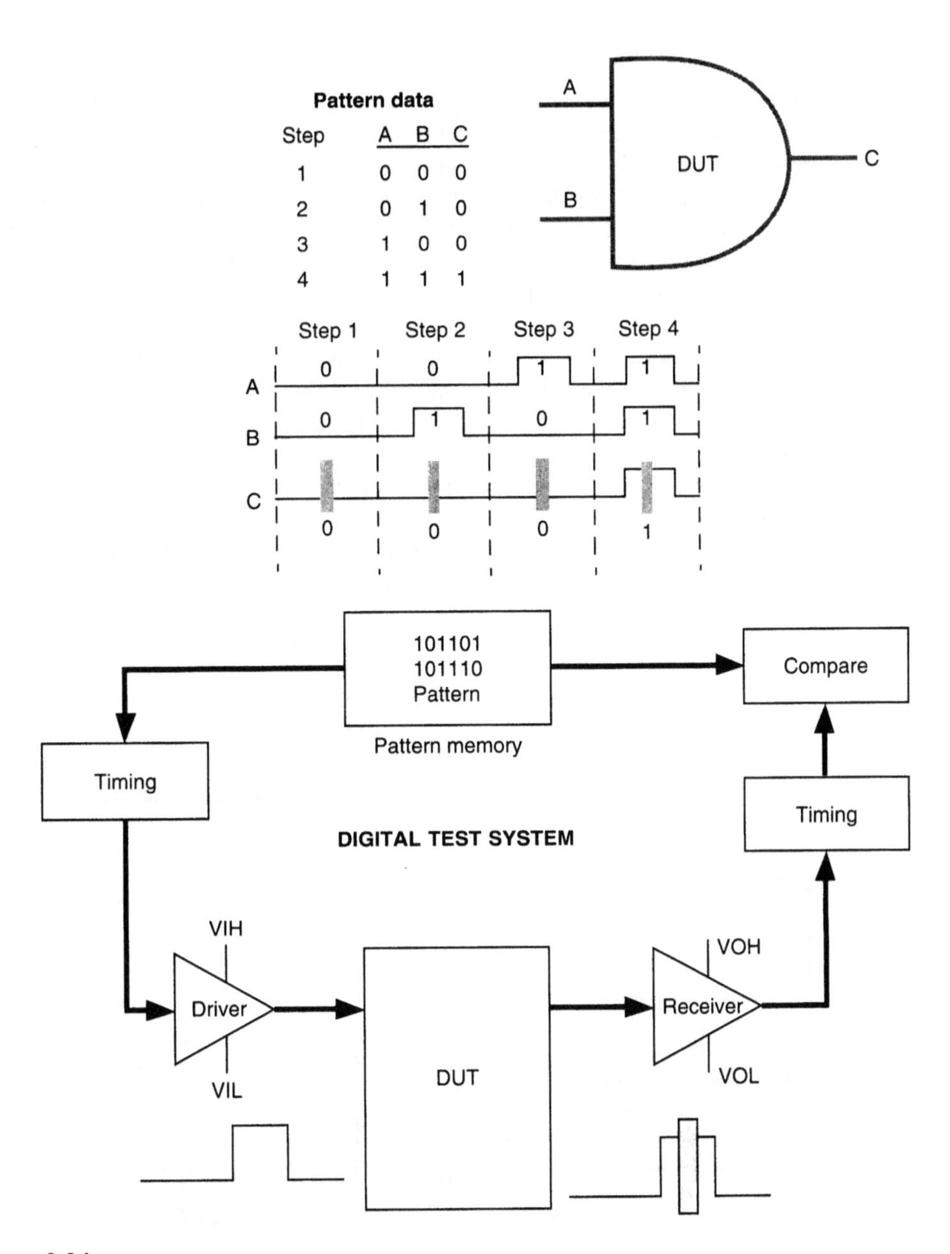

Figure 1.14

Logic Test System Overview

1.7.1 Analog-only Test System

The test system applies the analog signal to the device under test via the signal source. The fixed signal source instruments used by early analog test systems have been almost entirely replaced by programmable signal instruments using digital synthesis techniques. An analog test system will *source* analog data, *capture* analog data, and *analyze* the captured analog data in relation to specified signal characteristics. Instead of driving a pattern as a digital logic tester does, an analog source converts a numeric model of the wave shape into an analog signal.

The test system captures the output signal with an ADC instrument, and stores a numeric replica, or *digitized* representation, of the signal. The digitized signal is then analyzed for signal characteristics by the test system DSP. The signal capture system performs a function similar to the pin receivers on a digital system, with two important differences. First, the test system's analog capture instrument *digitizes* the DUT output with an ADC circuit, and stores a numeric replica of the signal information. Second, instead of comparing the device output with expected data, an analog test system performs an analysis of the output signal.

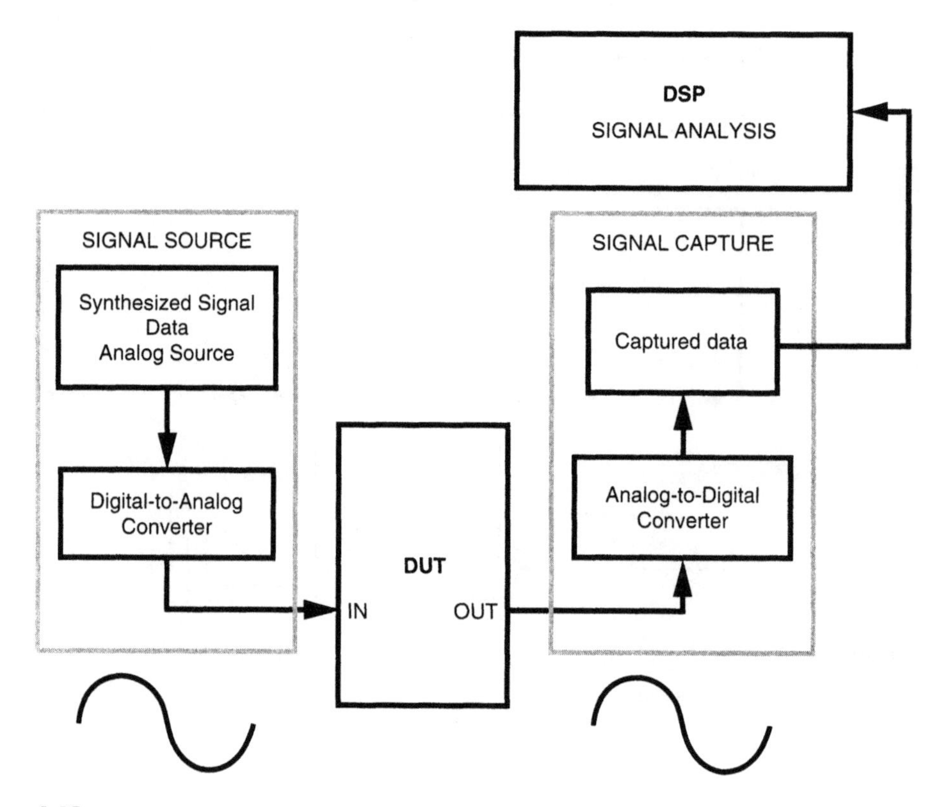

Figure 1.15

Analog Test System Overview

1.7.2 Mixed Signal Test Systems—Signal Source

On a mixed signal test system the signal source is an arbitrary waveform generator with both analog and digital outputs. If the device under test is an ADC, then the tester must supply an analog input signal in analog form. If the device under test is a DAC, then the test system must supply an analog input signal in digital form. A numeric model of the waveform, called a sample set, is stored in the source RAM. This sample set is repeatedly cycled out to the DAC of the analog section, or to the timing and formatting circuit of the digital output section.

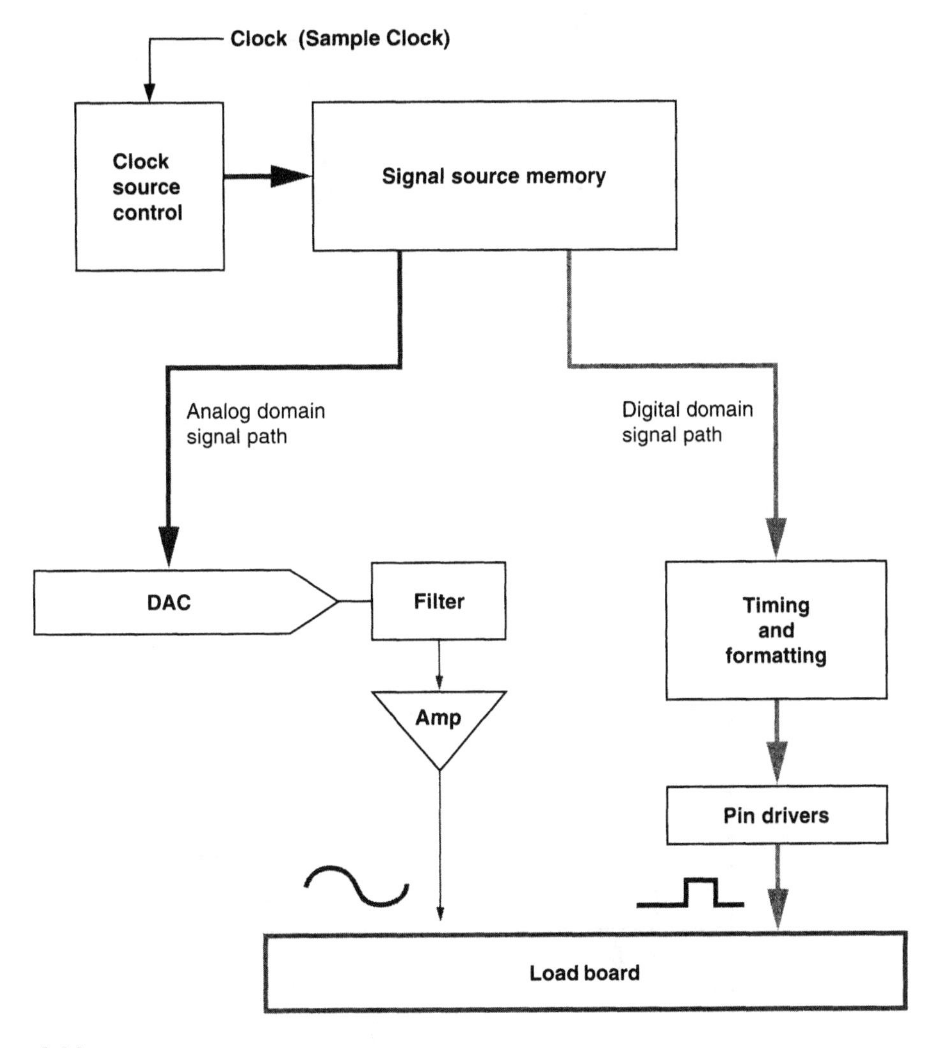

Figure 1.16

Signal Source Block Diagram

The *analog* side of the source works like a RAM behind a digital-to-analog converter. A numeric model of the waveform, called a sample set, is stored in the source RAM. This sample set is repeatedly cycled out to the DAC. The DAC, in turn, generates a sequence of voltage levels that produces the wave shape in analog form. The wave shape generated by the DAC is usually filtered and amplified before being presented to the DUT input.

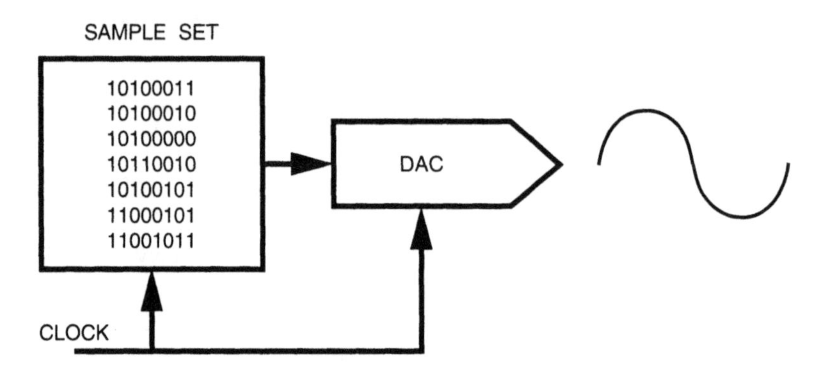

Figure 1.17

Analog Signal Source: RAM Behind a DAC

The same signal source data set can be presented to the device in digital form through the digital output path. In that case, the signal data set is processed sequentially through the test system's digital output timing, format, and drive circuits. The *digital* side of the source works just like drive data on a logic tester, except for *where* the data is stored, and *what* it represents. On a logic tester, the drive data is stored in pattern memory, and it represents the logic function input.

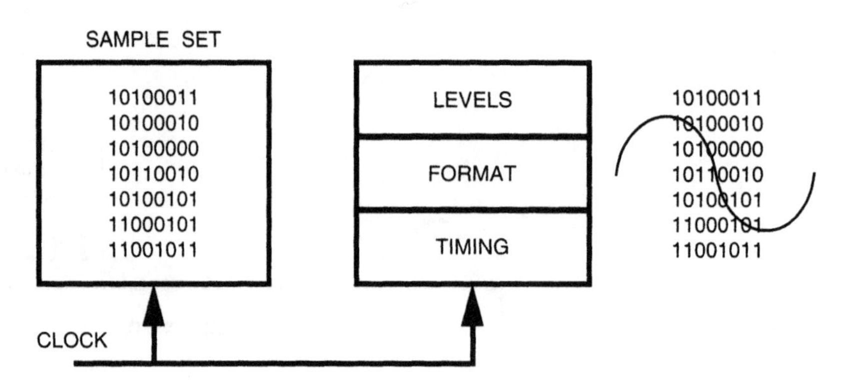

Figure 1.18

Digital Signal Source: Analog Data in Digital Form

On a mixed signal tester, the source data may be stored in pattern memory or dedicated source memory, and it represents signal information. You still need to worry about timing, formatting, and levels, but the data no longer comes from the pattern, and it does not represent a logic function.

1.8 More about the Signal Source

The hard part about getting the source to provide a proper input to the device is generating a periodic sample set. Remember that the source repeatedly loops on the sample set to generate a continuous wave form. The sample set, therefore, has to be set up so the loop-back generates a smooth transition.

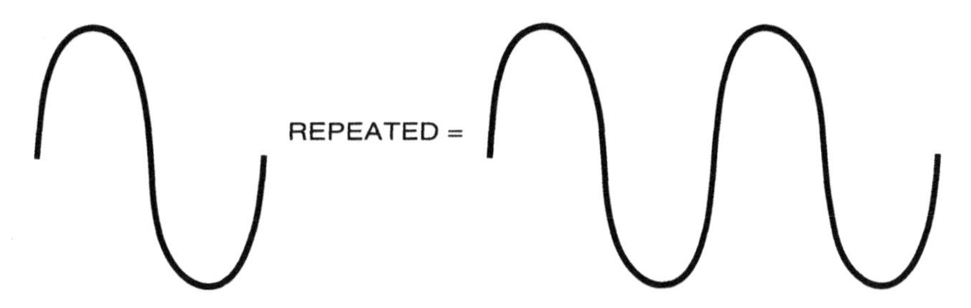

Figure 1.19

Sourcing a Periodic Sample Set

It's usually helpful to be able to get the points in the sample set close together—it makes the wave shape smoother. The faster we can clock data through the DAC, the more sample points can be generated within a given period.

Figure 1.20

Samples per Cycle

Amplitude resolution is also important. If we can only generate a few discrete levels, the output waveform will be crude. A 12-bit DAC, however, can generate 4,096 discrete levels, providing a higher level of accuracy.

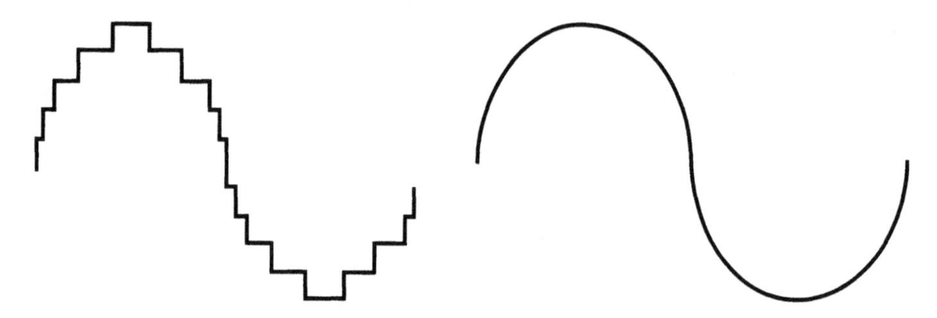

Figure 1.21

Amplitude Resolution

1.9 Mixed Signal Test System—Signal Capture

The capture section of a mixed signal tester is almost a mirror image of the signal source, acquiring analog information in either analog or digital form. Revisiting the previous example, let's say the device under test is a digital-to-analog converter. The output of a DAC, of course, is an analog signal. Conversely, if the device under test is an analog-to-digital converter, then the device output is in digital form, but it's not logic! The signal capture is a waveform digitizer with both analog and digital inputs.

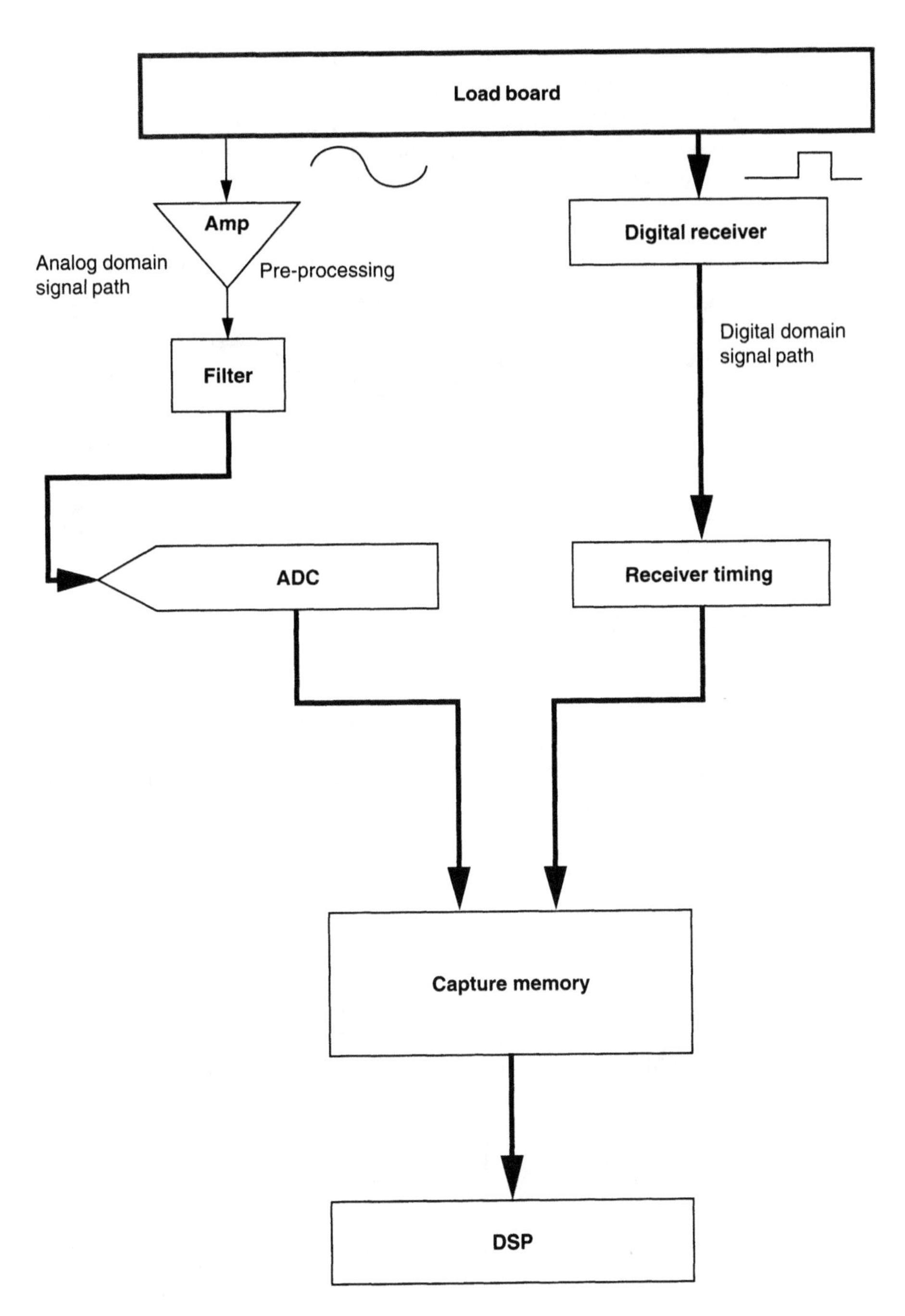

Figure 1.22

Signal Capture Block Diagram

The analog side of the capture works like an ADC driving a RAM. The ADC samples and converts the analog signal from the device into a digital code corresponding to a numeric value. Each sampled point on the input waveform is successively digitized and stored in the capture RAM, resulting in a numeric replica of the analog signal from the device.

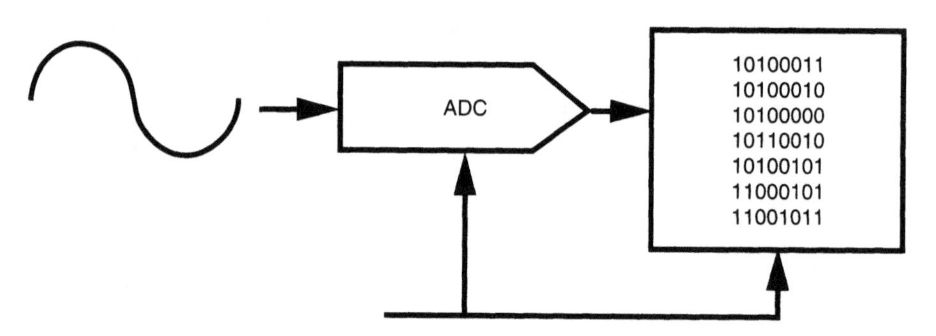

Figure 1.23

Analog Signal Capture: ADC Writing to RAM

Like the source, the digital side of the capture is similar in function to the receive data on a logic tester. On a logic tester, the expected data is stored in pattern memory, and it is compared with the strobed logic states from the digital receiver. On a mixed signal tester, there is no expected data. Digital values from the DUT are stored in capture RAM and analyzed as a signal.

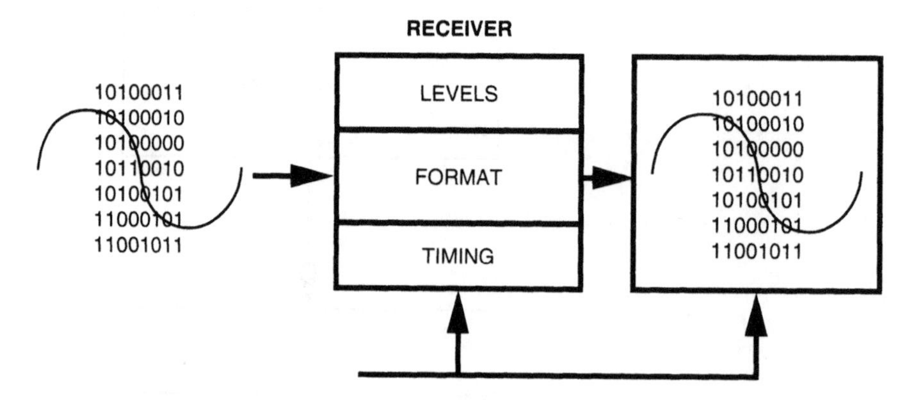

Figure 1.24

Digital Signal Capture: Analog Data in Digital Form

1.9.1 More about Signal Capture

The capture section faces some of the same problems as the source. Setting up a periodic sample set on the capture side means the test system must capture an integer number of cycles for the signal under test. Acquiring an incomplete number of cycles of a signal will cause the capture "snap-shot" to misrepresent the actual data.

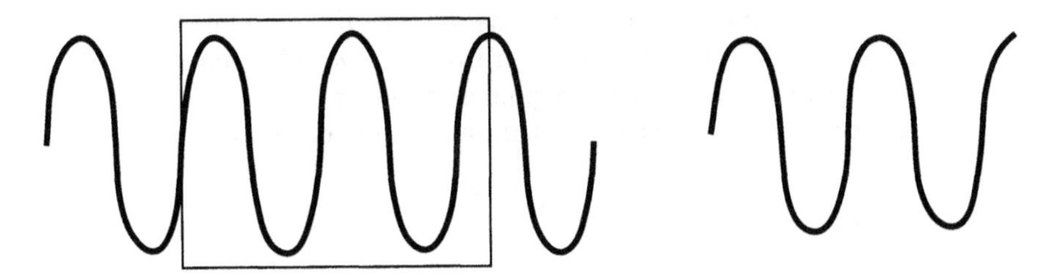

Figure 1.25

Capturing a Period Sample Set

Like the source, it's helpful to be able to get the points in the sample set close together—the faster we can clock data through the ADC, the less chance there is of "missing something" in the acquired data set.

Amplitude resolution controls the level of measurement precision. It's a lot like using a voltmeter—the more digits we have, the less the uncertainty.

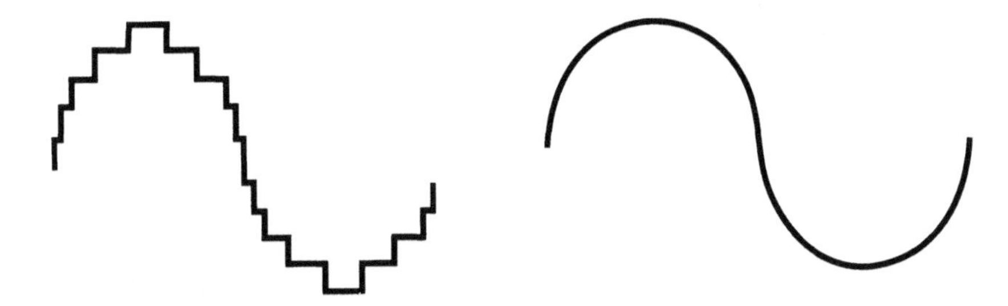

Figure 1.26

Adequate Capture Amplitude Resolution

1.10 Mixed Signal Test System—Signal Analysis

The **analysis** section of a mixed signal tester evaluates the captured data as a signal. A digital signal processor (DSP) is used to extract signal characteristics in both the time domain and frequency domain.

A DSP processes the data from Capture RAM. Rather than comparing each bit in each vector against expected data, the DSP processes the entire sample set *as a signal.* The operational element of a DSP system is an array.

The DSP in a test system can be programmed to perform measurements as a virtual instrument. The DSP system does not control measurement hardware. It *replaces* it. The sample set representing the device output signal is processed with various DSP algorithms in order to analyze the data set for signal characteristics.

Using the DSP is like writing a separate sub-program within the test program using specialized commands. The DSP can use data from Capture RAM, from the main test program, from another array, or from disk. Results from the DSP sub-program are passed back to the main program as parametric values. These parametric values are then compared against pass/fail limits in the main program.

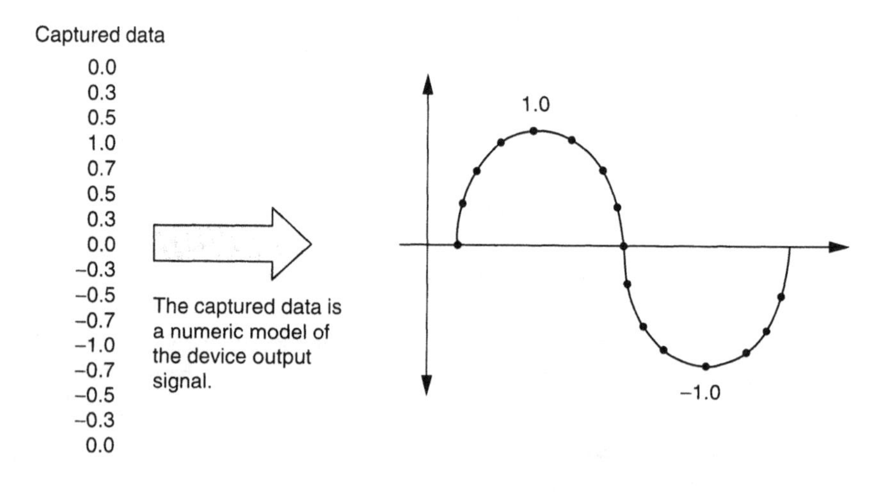

which is analyzed for signal characteristics

Harmonic Distortion

Signal to Noise

Figure 1.27

Signal Analysis

1.10.1 The Big Picture

A mixed signal test system will *source* analog data in either digital or analog form, *capture* analog data in either analog or digital form, and *analyze* the captured analog data in relation to specified signal characteristics.

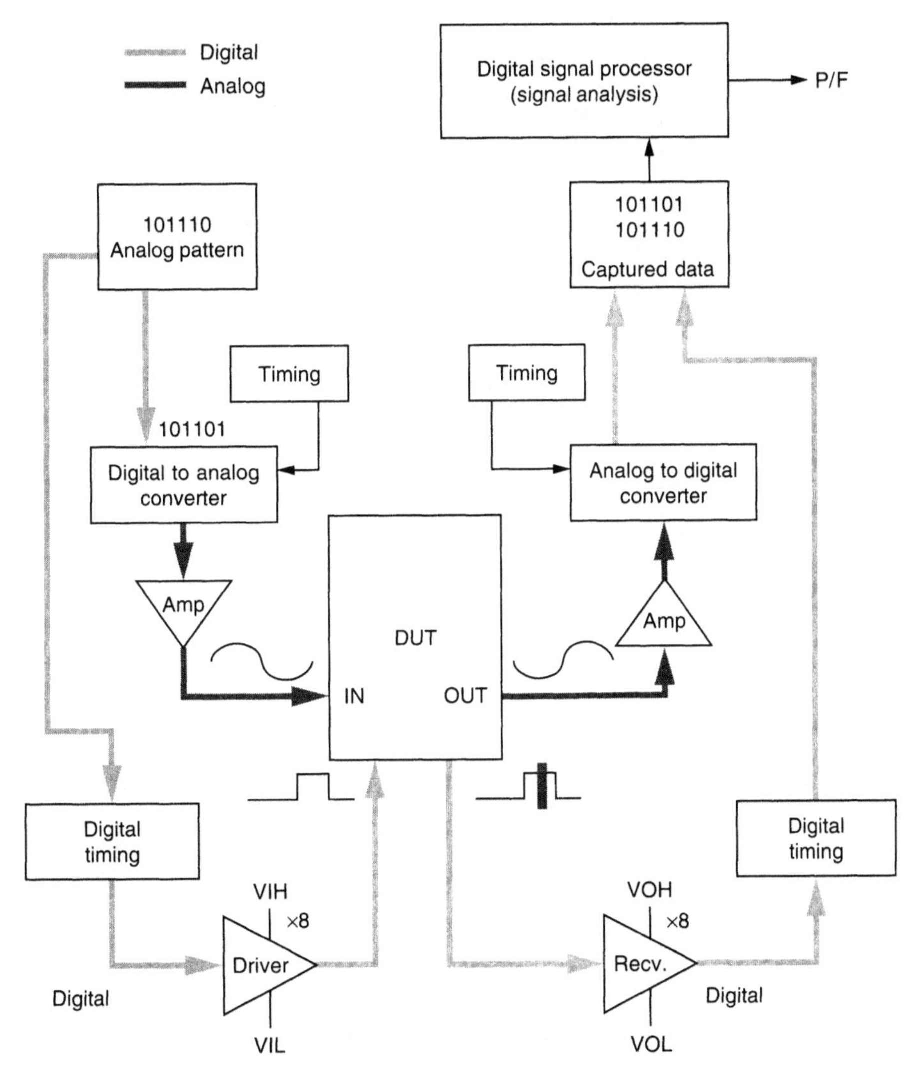

Figure 1.28

Mixed Signal Test System—Analog Instruments Only

Many mixed signal device types combine logic and mixed signal functions. The test system must *combine* logic test and mixed signal test on one platform.

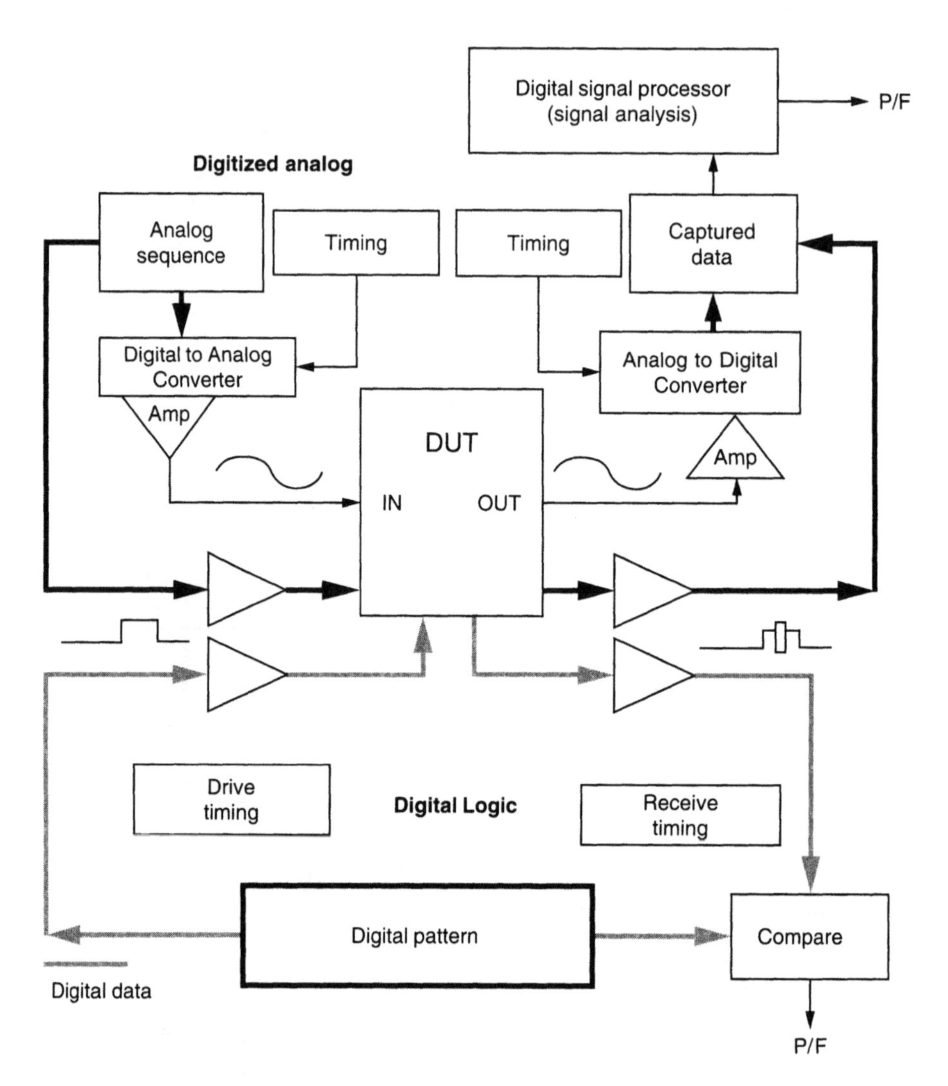

Figure 1.29

Mixed Signal Test System with Digital and Analog Instruments

1.10.2 More About Signal Analysis—Time Domain

Time domain signal analysis concerns the kind of signal information you might see on a oscilloscope.

- Peak Amplitude
- Area under the curve (RMS)
- Pulse Width
- Frequency
- Rise and Fall Time
- Overshoot and Undershoot
- Jitter

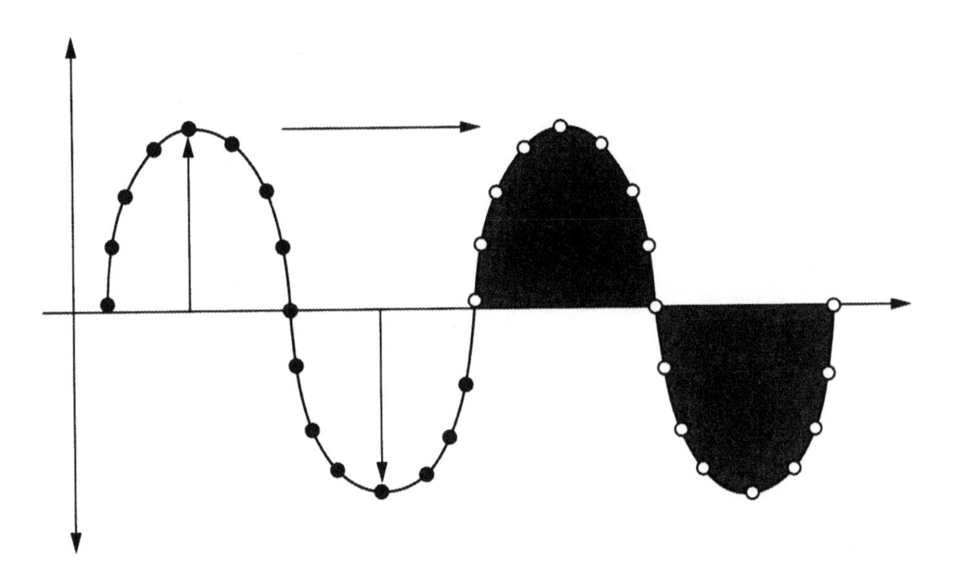

Figure 1.30

Time Domain Signal Analysis

1.10.3 Signal Analysis—Frequency Domain

We've all been around oscilloscopes enough to feel pretty comfortable with time domain measurements, but what in the world is frequency domain? As it turns out, frequency domain is the kind of stuff you see on a spectrum analyzer. Let's set the stage by looking at a special type of signal known as a multi-tone. In the example so far, we have been illustrating analog signals as a simple sine wave, as in Fig. 1.31.

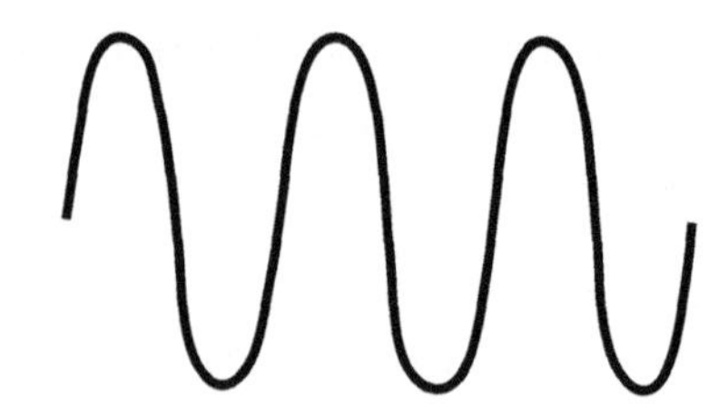

Figure 1.31
One Single Tone Sine Wave

Sometimes it is useful to have one signal that sums *several* signals into *one* waveform, as in Fig. 1.32.

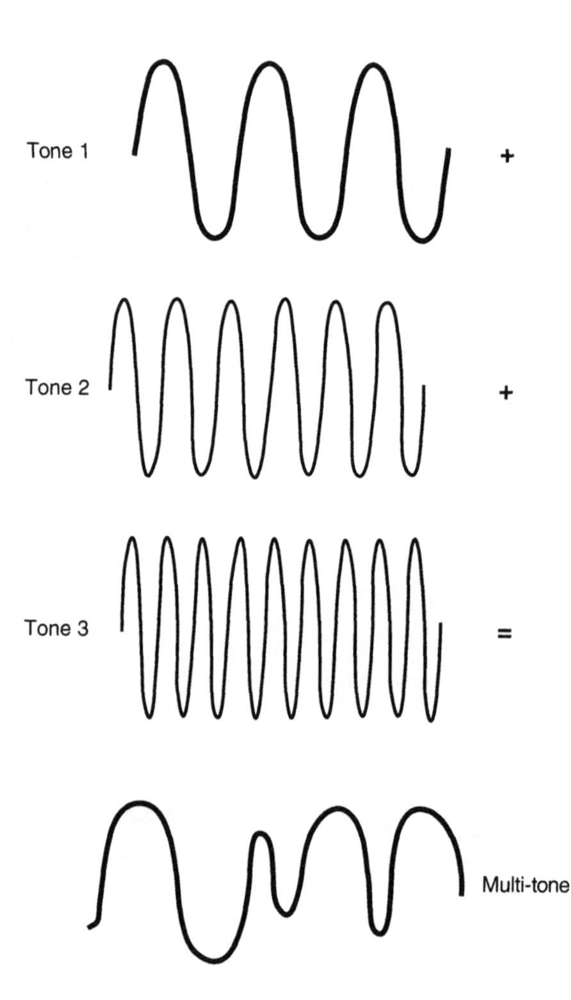

Figure 1.32
Multiple Single Tones

A multi-tone, therefore, is one signal that is made up of several different frequencies. Frequency domain analysis *reverses* this process. Transform functions deconstruct complex signals into their elemental components, and plot the relative amplitude of each component.

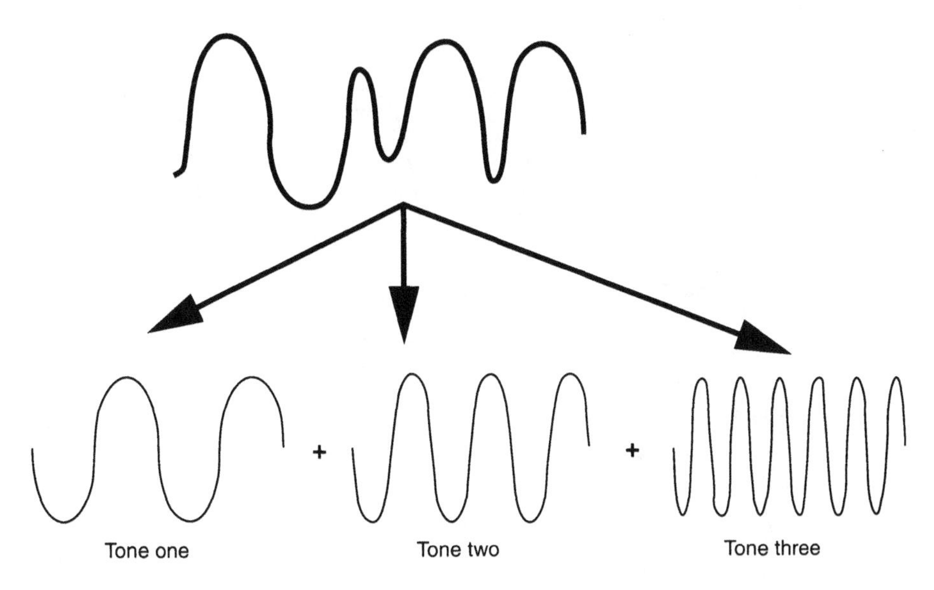

Figure 1.33

The Multi-tone Waveform Combines Several Sine Wave Data Sets

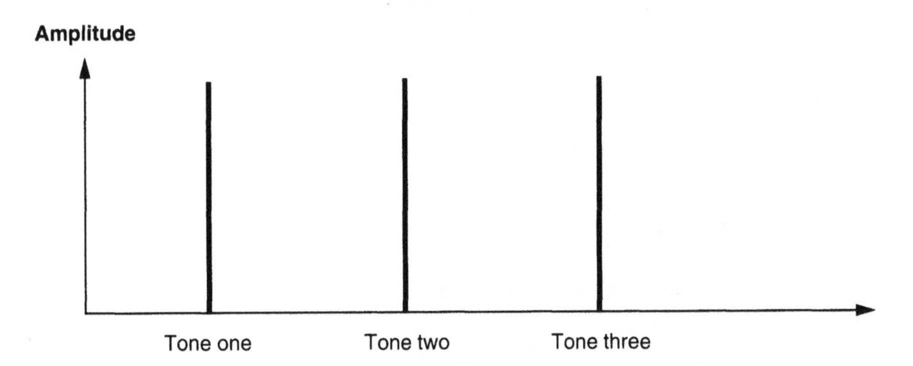

Figure 1.34

Frequency Domain Analysis—Amplitude Across Frequency

Once we've got frequency domain data, what can we do with it? Frequency domain testing often concerns evaluating characteristics of the signal shape or signal purity. Some features of the signal shape are more easily quantified in the frequency domain than in the time domain. For example, we might be able to use a scope to evaluate if a signal is noisy. But exactly how noisy is it? What is the amplitude and frequency of the noise? It's hard to tell in the time domain, but looking at the signal in the frequency domain makes it crystal-clear.

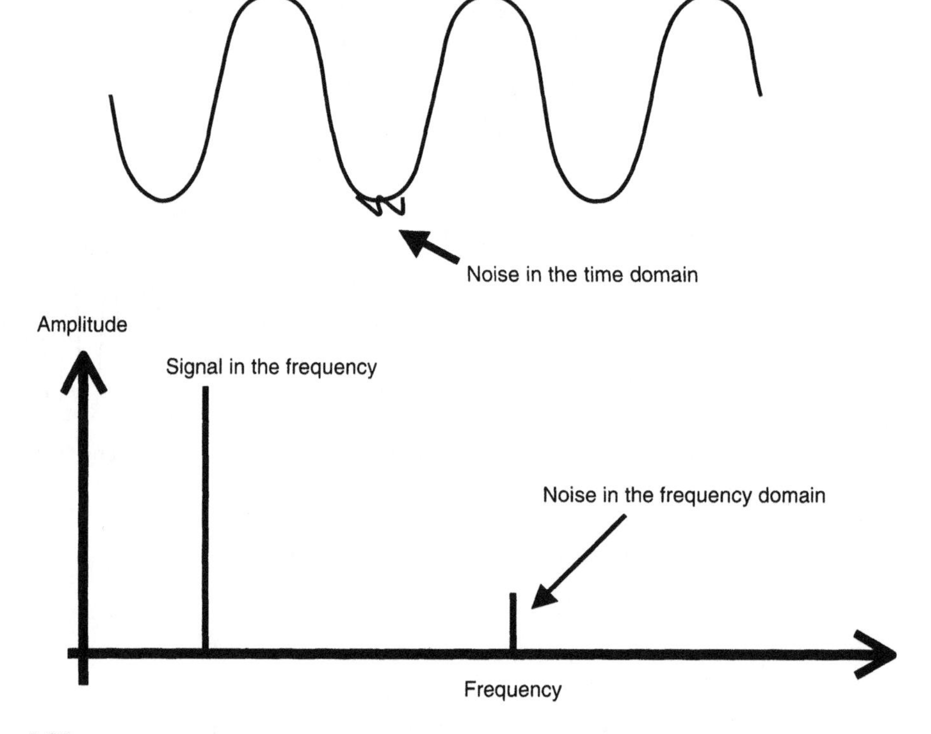

Figure 1.35

Measuring Noise in the Frequency Domain

Before we say goodnight . . .

Top Five Reasons to Know Mixed Signal Test

(with apologies to David Letterman!)

5. In the next industry recession, this knowledge will qualify us for jobs in TV repair and alarm installation.
4. Our children and spouses seem to know a lot about mixed signals.

3. Everything is analog, anyway; and returning to our roots is a great cure for mid-life crisis.
2. The digital test acronyms are getting old. It's time for some new acronyms And what *does* ISDN stand for, anyway? (ISDN stands for "**I** **S**till **D**on't **kn**ow!")
1. Maybe we can use Digital Signal Processing to predict the stock market.

Chapter Review Questions

1. Mixed Signal Test is: (*True* or *False*)

   a. Arbitrary, Subjective, and Asinine. T F

   b. The same as Logic Test T F

   c. Based on analyzing device data as a signal. T F

2. The three sections of a Mixed Signal Test System are

   a.

   b.

   c.

3. In Logic Test, the device performance is evaluated by comparing the device output with expected binary logic states. In mixed signal test, the device output is tested by (*choose one*)

   a. Reading poetry.

   b. Making up the answers.

   c. Evaluating the device output data *as a signal.*

CHAPTER 2

MIXED SIGNAL TEST MEASUREMENTS AND PARAMETERS

Wagner's music is not as bad as it sounds.

—Mark Twain

So, you're sitting in your office, pondering the spec sheet for a mixed signal device. The device does all kinds of cool stuff, but the trick is how to test the silly thing? About now, you might be thinking of a career change; or maybe joining the French Foreign Legion. Now what do you do? How can you determine if a mixed signal device is operating according to specifications? What are you going to measure? A large portion of mixed signal testing focuses on signal analysis. In this chapter, we will examine signal characteristics and their associated test and measurement techniques.

The procedure for signal analysis is simple: Apply the conditions, make some measurements, and perform some calculations. The signal analysis procedure usually produces a number, or set of numbers, that indicates some qualitative value corresponding to a signal characteristic. We're looking for signal characteristic values, or the results of the analysis, to be within a certain range. Like hand-grenades and horseshoes, as long as the signal characteristic is close enough, we'll say it passes the test.

If your background is in digital logic test, this may seem a little arbitrary. After all, the part is either good or bad; it either passes the test or it doesn't. Right? But the truth is even some tests for logic devices, such as IDD current, are specified as "close enough" tests. If the IDD specification is 100 mA, that doesn't mean that the device IDD must measure exactly 100.0 mA.[1]

What, or who, determines the specification for a signal parameter? (You're not going to like this.) Signal characteristic specifications are not based on absolute truth. If your device has a total harmonic distortion level of 0.2%, is that good or bad? It depends. Mostly on what the marketing department says! Read this table and weep.

Measured Value	Customer's Expectation	Competitor's Claim
0.2% THD	0.5% THD	1.0% THD
0.2% THD	0.1% THD	0.1% THD

Table 2.1

Market-Driven Specifications

A distortion level of 0.2% is good, if the customer wants 0.5% and your competitor can't deliver a part that is better than 1.0%. On the other hand, if the distortion measures 0.2%, the customer wants 0.1%, and your competitor delivers 0.1% devices at half the price, that's a bad part. Very bad! On the bright side, even if the specifications are not based on absolute truth and cast in eternal stone forever, the methods for making the measurements are deterministic. That's why they call it test *engineering*, after all. And yes, we'll show you how to do a harmonic distortion test, a little later.

Example

You want to test an amplifier circuit for DC gain.

1. You apply the condition of 50 mV volt on the input.
2. You measure the output and read 0.4965 volts.
3. You calculate the gain, as the ratio of input voltage to output voltage.

$$\text{Gain} = V_{out}/V_{in} = 0.4965/50 \text{ mV} = 9.93$$

[1] Sometimes, I wish I had never sent in that matchbook cover—you know the one I mean, "Learn Electronics! Make Big Money!!!" I probably should have gone for locksmith training, instead...

4. You compare the results against limits, to see if the gain is close enough. The maximum specified gain is 10.5, while the minimum specified gain is 9.5.

Minimum Gain: 9.5 Maximum Gain: 10.5 Results: 9.93

Evaluation: Close Enough!

2.1 Signal Analysis Categories

In general, signal analysis falls within one or more of three distinct categories—DC, time domain, and frequency domain. DC (direct current) signal analysis is used to determine the static or quiescent characteristics of the device, such as supply current or output pin voltage levels. Time domain, or AC, signal analysis applies to transient or dynamic signal characteristics. Typical time domain specifications include slew rate, pulse width, and settling time. The third category of signal analysis concerns measurements in frequency domain, such as noise and distortion. Mixed signal test employs a variety of methods to analyze the device response by measuring DC, time domain, and frequency domain characteristics.

2.2 Units of Measurement

Each signal analysis method has its own terminology and units of measurement. DC voltage measurements are expressed in volts or as a ratio, referenced to a specific test condition. The measurements may be either direct or inferred. Some voltage measurements are mathematically derived. Gain measurements, for example, are expressed as a ratio of output level to input level. The units may be in volts per volt, or in decibels. (We'll talk more about decibels a little later.)

DC current measurements are expressed in amps or as a ratio, referenced to a specific test condition. Test system hardware usually measures current indirectly, based on the measured voltage drop across a known resistance.

Time domain measurements express a rate of change in voltage or current, and are referenced to seconds. Frequency can be indirectly measured in the time domain, and is expressed in Hertz (Hz). Evaluating signal characteristics such as pulse width or duty cycles measures the elapsed time between signal level thresholds. Other time domain characteristics, such as slew rate, are expressed as a ratio of voltage and time. Signal energy is a more sophisticated measurement, derived as the root mean square (RMS) of the signal amplitude across time and expressed as volts RMS.

Some signal characteristics are more easily quantified through analysis in the frequency domain. Noise, for example, can be detected in the time domain but not precisely measured. The frequency domain, however, allows straightforward evaluation of signal characteristics including noise and distortion. Frequency domain measurements are typically relative values, expressed as a decibel ratio or as a percentage. Noise is an amplitude measurement expressed either in volts or in power, or as a ratio of the signal amplitude to the noise figure. In the context of this book, noise is defined as a spurious AC signal, not harmonically related to the reference signal, or as a non-periodic error. We will define distortion as an AC signal error that repeats for each period of the reference signal, hence, a periodic error. Distortion is an amplitude measurement, expressed either in volts or in power, or as a ratio.

2.3 Decibel Calculations

A decibel is a method for describing a ratio. The original unit of measure was defined as the Bel after Alexander Graham Bell. (The Bel unit was originally defined as a measure of relative acoustic power as perceived by the human ear, and the ear has a logarithmic response.) One Bel is a huge value that no one ever actually uses, like a one-farad capacitor. In practice, one-tenth of a Bel is used to describe voltage and power ratios—a deci-bel. The term "dB" is an abbreviation for decibel, which in turn is simply a method for expressing a ratio.

There are many different ways to describe ratios. For example, the ratio of one nickel to a dollar can be described as 1:20, which indicates that there are 20 nickels to the dollar. Or, you could say that one nickel is equal to five percent (5%) of a dollar. It's the same ratio, with different descriptions. The percentage references both values in the ratio to a scale of 100 so a ratio of 1:100 is the same as 1%.

Table 2.2

Ratio Representations

Ratio	Percentage
1:1	100%
1:10	10%
1:100	1%

Like a percentage, the decibel (dB) is a way of describing a ratio. Instead of referencing the ratio to a scale of 100, the dB system uses a *logarithmic* scale of log base 10. (When we see the expression log(x), it does not mean log times x, it means the log10 of x.) When used to describe a voltage ratio, a decibel ratio is defined as the following equation:

$$dB = 20 \times \log\left(\frac{v1}{v2}\right)$$

where $v1$ and $v2$ are the two voltage levels, or components, of the ratio.

The expression derives the log of the ratio of $v1$ to $v2$, and we use a scale of 20 for measuring voltage, or 20 times log (x). There is a direct correlation between dB and percentage. Here's the key: one percent is the same as 40 dB. Increasing the percentage ratio by a factor of 10 correlates to a 20 dB ratio increase.

Table 2.3

Correlating dB Levels with Percentages

Percentage	dB
1%	–40 dB
0.1%	–60 dB
0.01%	–80 dB
0.001%	–100 dB

2.3.1 Negative dB

Because a smaller ratio produces a larger dB value, the conventional representation of dB ratios are calculated "upside down." The dB ratio of 1:1, or 100%, is equal to 0 dB. A ratio that is less than 100% logically correlates to a dB ratio that is less than 0 dB, or a negative number. A common method of referencing the calculation to 0 dB is to use a scalar of negative 20 (–20).

$$\text{Voltage dB} = -20 \times \log\left(\frac{v1}{v2}\right)$$

Where $v1$ is the reference level, and usually the larger of the two voltage levels.

Expressing the dB calculation as the ratio of the larger reference value to the smaller value has several advantages. First, the equation is mathematically consistent with the grammatical expression. The term "signal to noise ratio," for example, implies that the ratio is derived from the signal amplitude divided by the noise amplitude. The second advantage concerns computational accuracy. In the case of extremely small numbers, the ratio can be corrupted by round-off error if the smaller value is divided by the larger value.

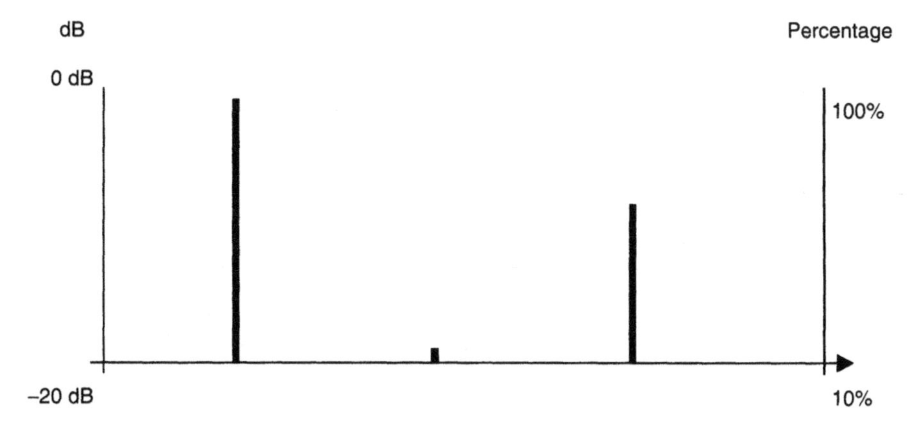

Figure 2.1

Example Percentage and dB Graph

2.3.2 Power dB Ratio

The term power has some special considerations when calculating the dB ratio. Mixed signal tests are often concerned about the actual shape of the signal, and the term power is used to describe the effective energy, approximately correlating to the area under the curve. (It's not exactly the same as area under the curve, but it's close enough. After all, this is mixed signal we're talking about here.)

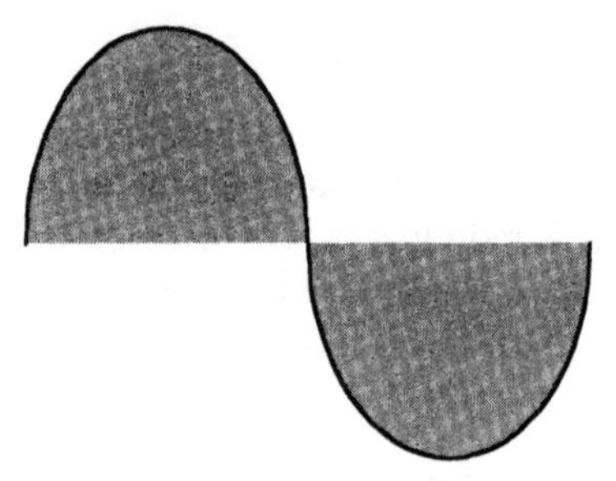

Figure 2.2

Calculating Relative Power

Reviewing the simple equation for power, we find that relative power is a function of the square of the voltage.

1. $Power = E \times I$
2. $I = \frac{E}{R}$
3. $Power = E \times \frac{E}{R}$
4. $Power = \frac{E^2}{R}$

Because log calculations are based on the law of squares, the effective squaring of the voltage level must be taken into account when calculating power ratios. To properly associate the power and voltage ratios, the dB calculation for power uses a scale of 10.

$$\text{power dB} = -10 \times \log\left(\frac{p1}{p2}\right)$$

where $p1$ and $p2$ are the relative power levels.

You don't believe me, do you? That's cool, logs are squirrelly little buggers. Let's look at it this way. The log of 5 is about 0.6987. Try it on your calculator, and see if you get the same answer. (If you don't, you are probably using ln, the natural log, instead of log as in logarithm.) OK, now let's see what happens with 5 squared. If you calculate the log of 25, you get 1.39794. Now, watch carefully as we pull the rabbit out of the hat: Voila! 1.39794 is equal to twice the value of 0.6997.

The same thing is going on with power and voltage. If you've got 5 volts into 1 ohm, then the power is equal to 25 watts.

Voltage dB = 20 × log (5 volts) = 13.979 dB

Power dB = 10 × log (25 watts) = 13.979 dB

Is that cool, or what?

2.4 Signal Analysis and Test Methods

Signal analysis is the last step in the generalized test process of source, capture, and analysis. The first part of the test process establishes the correct conditions, or stimuli. Signal information from the device under test is captured, and then evaluated according to specifications. We will examine the three aspects of signal analysis—DC, time domain, and frequency domain—in the context of a test sequence for mixed signal device. The example device is a programmable gain amplifier (PGA) with eight digitally programmed gain settings.

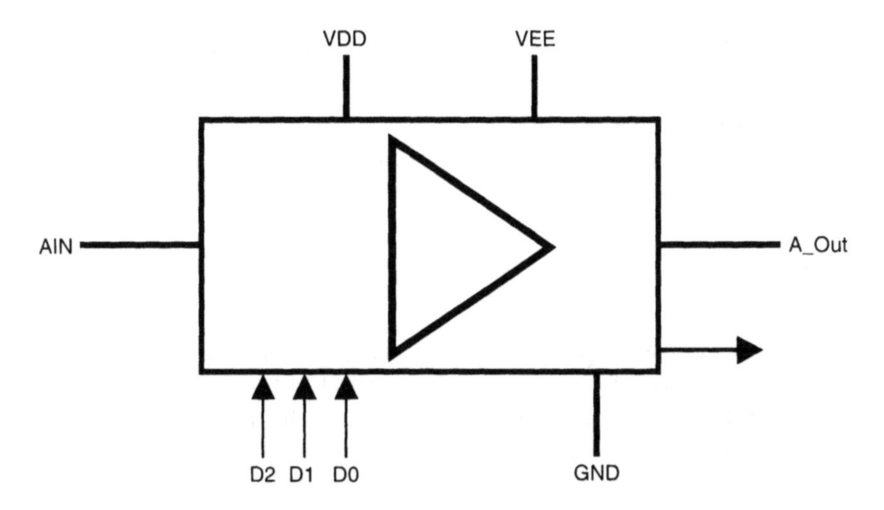

Figure 2.3

Mixed Signal Device Example

Pin Function

Pin AIN is the analog input to the amplifier.
Pin A_Out is the output of the amplifier.

Pins D0, D1, and D2 are the gain set digital inputs. D0 is the LSB, D2 is the MSB.

000	Gain = 1	100	Gain = 5
001	Gain = 2	101	Gain = 6
010	Gain = 3	110	Gain = 7
011	Gain = 4	111	Gain = 8

Pin OVR_Range is the over-limit digital output. This pin goes to a logic high when the output level exceeds 10 volts.
VDD is the positive voltage supply. Nominal VDD level is 12.0 Volts.
VEE is the negative voltage supply. Nominal VEE level is −12.0 Volts.
GND is the device ground pin.

2.4.1 The Test Plan

The test engineer creates a test plan document based on the device specification document or "spec sheet." The device specification document describes the operation and electrical characteristics of the device. In addition to the specification document, the test engineer also takes into consideration his or her own experience, knowledge of the ATE system capabilities, and prescribed conventions.

Designing the Test List

The order of tests is arranged in a sequence that will most quickly identify possible defects. The most basic tests are usually performed first, with the view that if the part fails the basic tests, then it is not necessary to test it any further. The order of tests must also take into consideration how a failure is identified and recorded in the data log.

For example, suppose the device under test has a defect in the power supply pin connection—the bond wire was not attached from the package pin to the die pad. If the first test is an output amplitude test, the device will fail and the failure will be recorded as a functional failure. Describing this device as a functional failure does not accurately identify the actual problem.

Some conventions have developed that help to organize the test flow to provide the most accurate information.

1. Is the tester connected to the device? (More on this later!)
2. Is the power supply current within spec? (IDD)
   (If not, there's no point in going further)
3. Is the input pin current within spec? (IIH, IIL)
   (If the inputs do not work, nothing else will work)

4. If the connections, the power supply, and the inputs are within specification, then the test flow can evaluate the device functionality.

The functional tests, in turn, also follow a logical order to properly identify the failure mechanism while also minimizing test time.

- Does the device perform the correct operational function? (If the part does not function, the output levels cannot be tested.)
- Can the output pin generate the correct signal with the specified current load?
- Are the measured time domain and frequency domain parameters within specification?

2.4.2 The Test List

Beginning with the specifications, the test engineer develops the test plan and test list.

Example Device DC Specifications

Analog Input Pins

Leakage = +/–1 uA at 10 volts input
Input offset voltage = +/–2 mV

Analog Output Pins

Maximum output voltage = +/–10.5 volts
Minimum positive output current = 5 mA at +10 volts
Minimum negative output current = –5 mA at –10 volts
Gain Error <2%
Linearity Error <1%

Digital Input Pins (Gain Set Pins)

IIL = +/–1 uA at 0.0 volts
IIH = +/–1 uA at 5.0 volts
VIL = 0.2 volts
VIH = 2.4 volts

Digital Output Pins (OVR_Range)

IOL = 5 mA
IOH = –5 mA
VOL = 0.2 volts

VOH = 3.2 volts
Threshold = 10 volts, +/–0.1 volts

Example Test List

Continuity: The continuity tests verify proper connection of the DUT. If the DUT is not present or not properly connected to the test system, no tests can be performed.

Supply Current: The supply current tests verify that the amount of current on the device power supplies is within specification. This test also is a way of checking for certain types of gross process errors. If there is a gross process error that causes a large supply current, it is more efficient to identify the flaw early in the test process.

Leakage Current: The leakage tests measure the amount of current flowing on the device input pins. Excessive current can cause unreliable operation. Measuring leakage current is another check for gross process errors.

Offset Voltage: The offset voltage test for analog amplifiers measures any required adjustment on the amplifier input to force the amplifier output to zero volts. The offset measurement combined with the maximum output level test is a quick verification of device functionality.

Maximum Analog Output: Leakage and offset tests verify proper operation of the device input circuits. The compliance of the output stage is verified by measuring the maximum output voltage level under a specified current load.

Over Level Function: One of the conditions for testing the over level function is identical to the condition for the maximum analog output test, so for the sake of efficiency the two tests are grouped together. If the amplifier output is at the maximum level, the OV_LVL pin should be active.

Gain Error: The gain test evaluates the overall span of the amplifier output. Once the minimum and maximum range of the amplifier is tested via the offset and maximum output level tests, the actual output span can be calculated and compared with the ideal.

Linearity Error: If the device is perfectly linear, each gain setting should cause the amplifier output level to increment in equally spaced steps. To verify acceptable linearity, the amplifier output is measured under various combinations of input levels and gain settings. The response of the device is compared with a calculated ideal, and variations from the ideal are identified as linearity errors. Linearity tests are more time-consuming than other tests, and therefore placed at the end of the test list. Only devices that have passed all other tests are candidates for additional test time investment.

2.5 DC Test Outline

2.5.1 Continuity Tests

The purpose of continuity tests is to verify that the test system is properly connected to the Device Under Test (DUT). Continuity tests verify that all DUT signal pins are connected to the tester channels. Continuity tests also verify that the pins of the DUT are properly connected to the internal device circuitry. The measurement results can reveal significant information on the DUT itself, in addition to verifying the DUT-to-tester connection. The test measurements may be used in a database for controlling and monitoring the manufacturing process. The continuity test is a powerful and simple test and yet it is often the most misunderstood. Continuity tests *do not* check a specified device parameter, and are not specified in the data sheet.

Continuity tests verify the presence of the internal diodes for each input and output pin. The internal diodes are tested by *forcing a current* and measuring the voltage drop. Because there are usually two diodes per signal pin, at least two tests are performed. By forcing a forward current through the upper or lower diodes, one expects to measure a diode drop, typically between 600 mV and 700 mV. Because the return path of the forced current is through VDD or VSS, VDD must be grounded.

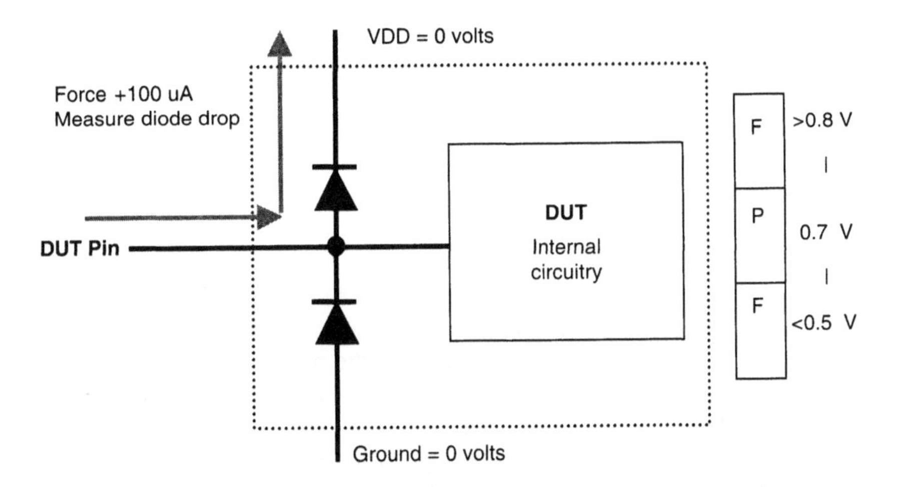

Figure 2.4

Testing the VDD Diode for Continuity

The VDD diode, or upper diode, connection for each pin is tested by forcing a *positive* current to activate the VDD diode, given that the device VDD pin is set to 0 volts. The current force resource is typically set to clamp at 1.0 volt. If the measured voltage drop across the diode is too high (typically greater than 0.8 volts), the diode connection is an open circuit. If the measured voltage drop across the diode is too low (typically less than 0.5 volts), the diode connection is shorted to VDD. The results are why continuity tests are sometimes called "opens and shorts" tests.

The VSS (GROUND) diode connection for each pin is tested by forcing a *negative* current to activate the VSS diode. The current force resource is typically set to clamp at −1.0 volt. If the measured voltage drop across the diode is too high (more negative than −0.8 volts), the diode connection is an open circuit. If the measured voltage drop across the diode is too low (less negative than −0.5 volts), the diode connection is shorted to VSS.

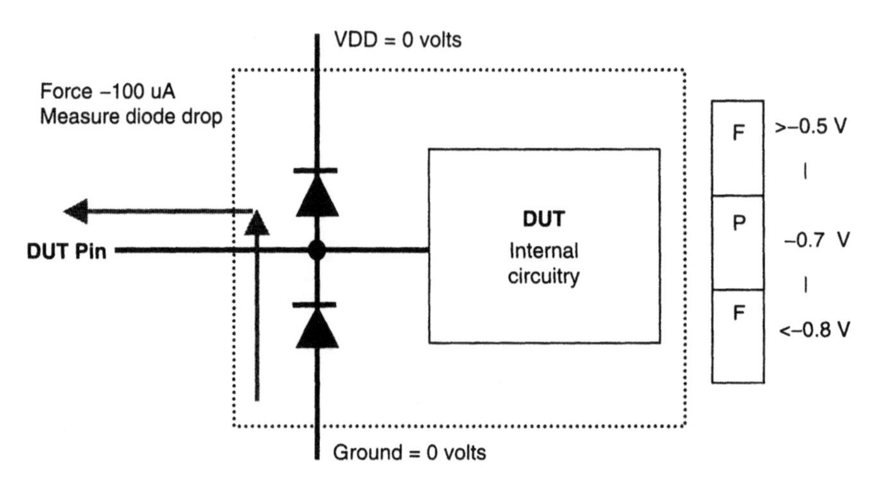

Figure 2.5

Testing the VSS Diode for Continuity

If VDD or VSS pins are open, or if the tested pin is shorted to VDD/VSS, the continuity test fails. This test does not detect if the pin under test is shorted to another pin. One method to detect pin-to-pin shorts is to perform the continuity test on each pin serially (one pin at a time), and to force all pins, except the pin under test, to 0.0 volts. (Forcing all pins to 0.0 volts is easily accomplished by using the test system functional drivers.) If the pin under test is shorted to another pin, the test will measure 0.0 volts instead of a diode drop voltage. The 0.0-volt measurement will cause the test to fail.

2.5.2 Supply Current Tests

The supply current tests verify that the DUT supply current is not excessive. Although it is usually not specified, it is sometimes good practice to check for a minimum supply current. There are two methods for testing the device supply current.

The first method is called static testing, because the device is not active. The second method is called dynamic testing, because the device is active while the current is being measured. Because the instruments for measuring DC values are slow in comparison to typical device execution speeds, dynamic testing usually makes use of a functional test loop. The device runs the same sequence repeatedly until the DC measurement is complete.

The IDD current can be measured once the device is in the specified condition. It is usually good practice to plan for a settling time delay *after* the conditions are programmed and *before* making the measurement. There are some factors that may prevent the device from reaching the specified condition immediately, particularly the load board bypass capacitors and the settling time of the ATE system instruments.

Supply Current Test Sequence

Force all input pins to 0 volts
Force all output pins to 0 mA
Force VDD to +12.0 volts
Force VEE to −12.0 volts

Wait for the ATE instruments and DUT to settle
Measure IDD current and compare with limits
Measure IEE current and compare with limits

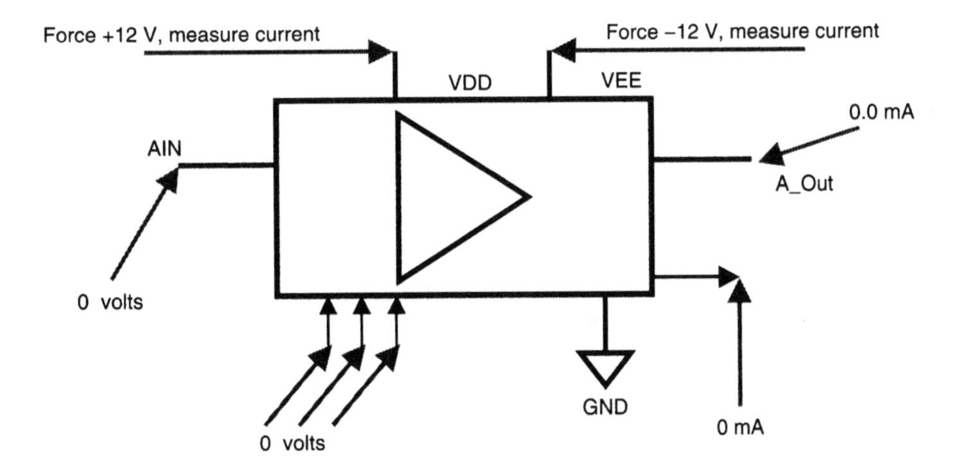

Figure 2.6

Power Supply Current Tests

2.5.3 Input Pin Current Tests (Leakage)

Input pin current tests verify that the device inputs do not require excessive drive current. Leakage tests are performed with the power supply pins set to the nominal operating level. IIL (input current at logic low) is tested by applying a logic state using the specified VIL level, and measuring the current flow into the pin.

IIH (input current at logic high) is tested by applying a logic state using the specified VIH level, and measuring the current flow into the pin.

To test for pin-to-pin leakage, it is common practice to pre-set the voltage level of all input pins to the opposite extreme of the pin under test. If the IIL test requires an input level of 0 volts, only the pin under test would be forced to 0 volts. The other input pins would be forced to a logic high level.

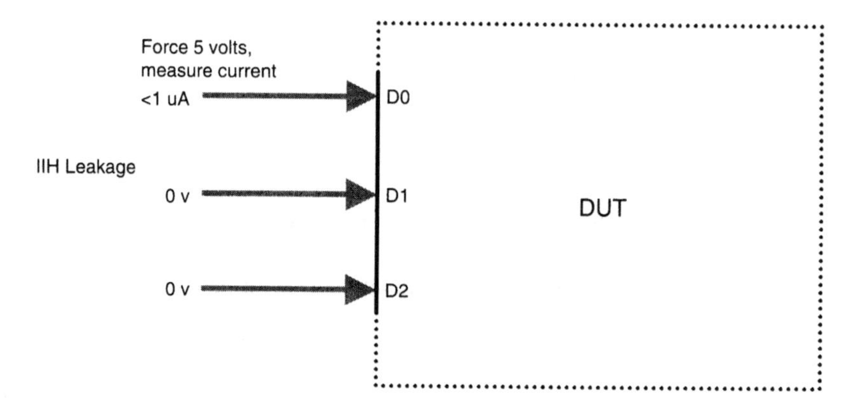

Figure 2.7

Input Current Test at Logic High (IIH)

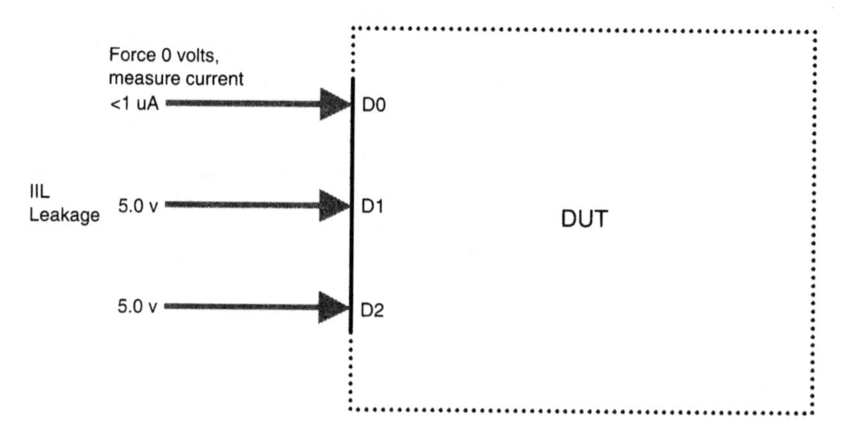

Figure 2.8

Input Current Test at Logic Low (IIL)

2.5.4 Offset Voltage

Offset voltage measures the voltage correction required on the amplifier input to force the amplifier output to zero volts. Because of process variations and imbalances in the internal circuitry, a zero volt level on the amplifier input does not always cause the amplifier output to generate a zero voltage level. In that case, the input must be adjusted to achieve a zero voltage output level. The amount of required adjustment or correction is the input offset. The device specification defines an acceptable range of offset values.

For most tests, the general approach is to apply a known input condition, and verify the output response. Offset tests reverse this approach. The objective is to determine the *input* level that corresponds to a known level on the *output*.

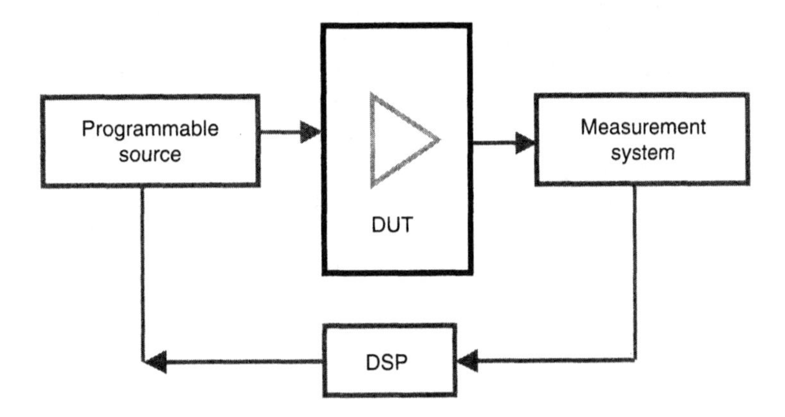

Figure 2.9

Adjusting and Measuring Input Level for Offset Test

In this example offset test circuit, the ATE system DSP unit controls the device input level via the Programmable Source instrument. By a successive approximation process, the DSP evaluates the DUT output level acquired by the measurement system, and adjusts the input level until the DUT output is zero volts. The input level required to force the DUT output to zero volts is evaluated as the input offset voltage.

2.5.5 Output Compliance Tests

Conductive parameter tests verify the drive capability of the amplifier output pin. The conductive parameter tests measure the output voltage level with a specified current load. Output current tests measure the current capacity on the output pin of the device when the output level is at the specified condition. Output voltage tests measure the voltage drive level on the output pin of the device for a specified logic state. The output voltage drive level is tested by verifying that the amplifier output can generate an acceptable voltage level with a specified current load.

Conductive tests verify that the voltage and current drive capability of the output pin under test is adequate. The device must be able to generate output levels with enough current to drive the circuit load. In the end-use application, a device that cannot supply sufficient current on the amplifier output pins will cause unreliable operation. The compliance of the DUT output circuitry is verified by measuring the maximum output voltage level under a specified current load. If the output is measured only under low current conditions, excessive "on resistance" of the output stage could remain undetected.

The device output levels are specified in conjunction with a current load. The Current Output Low (IOL) specification describes how much current the output must supply when generating a negative voltage level. IOL is referred to as device sink current, because current flows into the device toward ground. A negative output level "pulls down," so the tester resource must source current. The Current Output High (IOH) specification describes how much current the output must supply when driving a logic high. IOH is referred to as device source current, because current flows from the device toward ground. A positive output level "pulls up" with a positive current flow, so the tester resource must sink current.

The analog output of the DUT amplifier is rated at positive 10 volts with a 5 mA load, and negative 10 volts with a –5 mA load. The absolute maximum output level is +/–10.5 volts. To test the maximum output level, the device is powered with +/–12 volts, with all digital inputs set to 0 volts. The analog input is driven with 10.5 volts, and the analog output level is measured and evaluated against limits. The process is repeated using a –10.5 volt input to test the negative voltage output compliance.

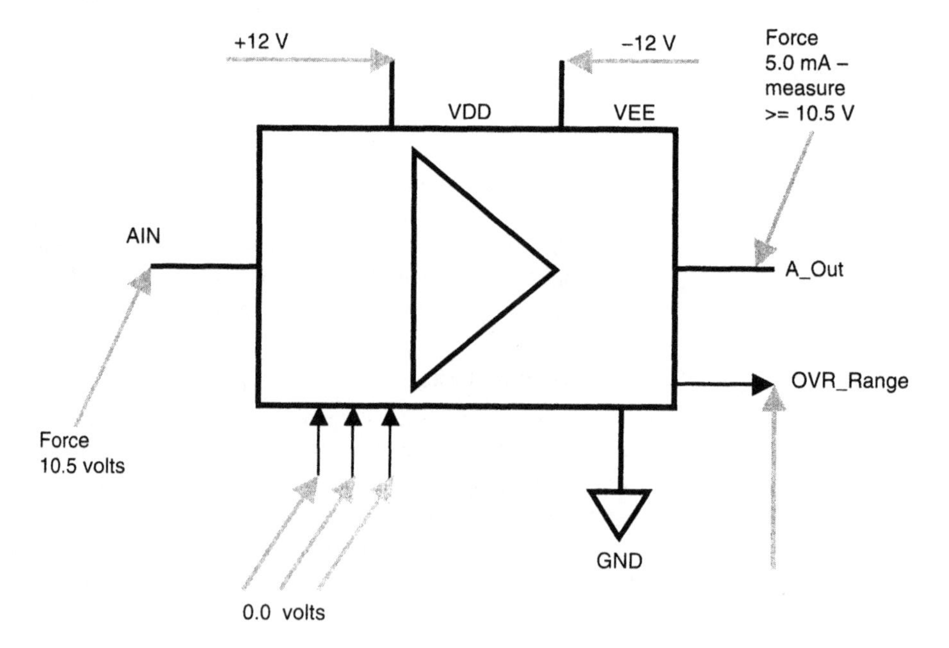

Figure 2.10

Testing Output Voltage Compliance

2.5.6 Over-Range Function

The device over-range function provides an indication of an over-range condition on the amplifier output via a logic level on the OVR_Range digital output pin. This pin goes to a logic high when the analog output level exceeds +/–10 volts, with a 100 mV margin. The maximum analog output test procedure has already verified that the OVR_Range pin goes to a logic high (>3.2 volts with a –5 mA load).

Another set of tests must verify that the over-range function produces a logic low when the analog output level is less than the maximum. The current load for the OVR_Range pin is set to +5 mA, and the logic threshold (VOL) is set to 0.2 volts. The analog input is set to +9.9 volts, and the OVR_Range pin is checked for a logic low. The analog input is set to –9.9 volts, and the OVR_Range pin is again checked for a logic low.

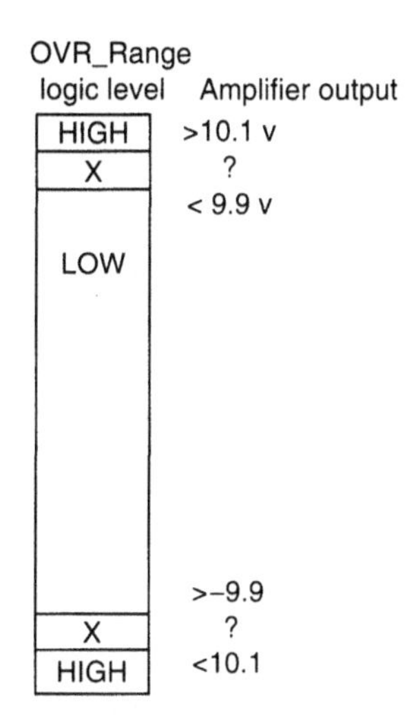

Figure 2.11

Testing Over-range Function

2.5.7 Gain Error Tests

The gain test evaluates the overall span of the amplifier output. Once the minimum and maximum range of the amplifier is tested via the offset and maximum output level tests, the actual output span can be calculated and compared with the ideal. Gain error is a measure of the overall device range, compared to an ideal range.

Ideal:

Gain Setting = 1 Input Level = 1.0 volts Output Level = 1.0 volts
Gain Setting = 8 Input Level = 1.0 volts Output Level = 8.0 volts
Ideal Output Level Span with a 1.0 volt input = 7.0 volts

Actual:

Gain Setting = 1 Input Level = 1.0 volts Output Level = 1.05 volts
Gain Setting = 8 Input Level = 1.0 volts Output Level = 8.15 volts
Actual Output Level Span with a 1.0 volt input = 7.1 volts

The difference between the actual and ideal is calculated as a percentage, as follows:

$$\frac{\text{Actual} - \text{Ideal}}{\text{Ideal}} \rightarrow \frac{7.1\ v - 7.0\ v}{7.0\ v} = 0.0142 \times 100 = 1.42\%$$

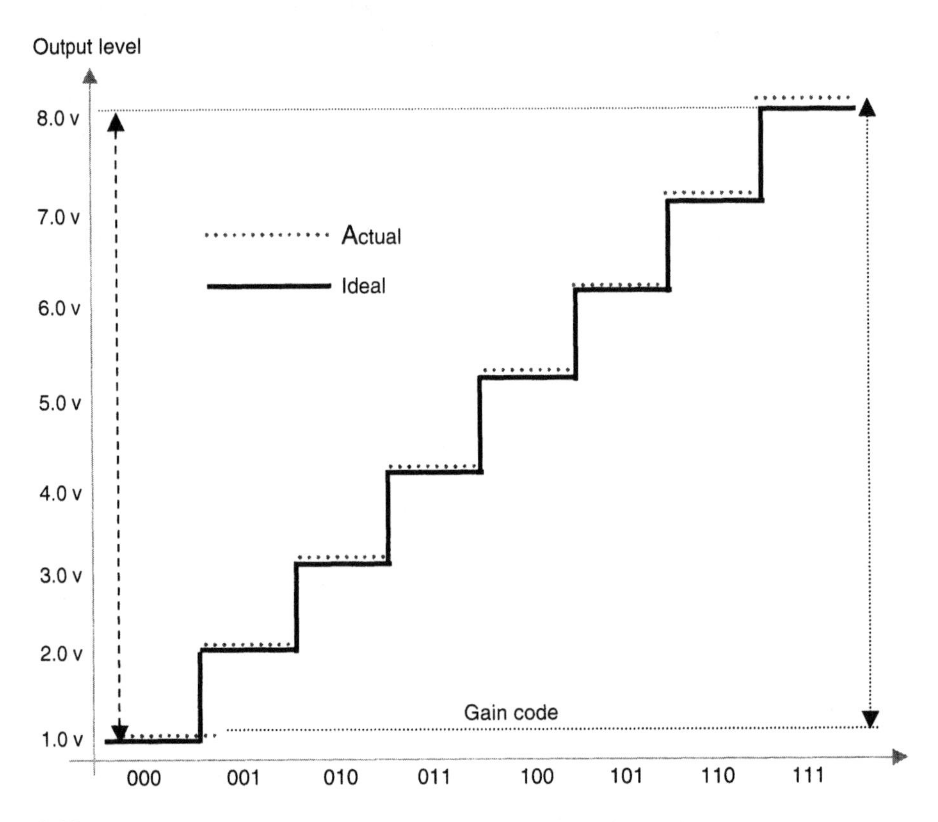

Figure 2.12

Gain Testing Measures Overall Span

2.5.8 Linearity Error Tests

Linearity error measures each gain step by changing the gain setting with a constant input voltage level. The incremental steps of the output are compared to a calculated linear "straight line." Based on the two measured end points derived from the gain test, we can calculate the overall span for this device as 7.1 volts. Between the first and last level, there are 7 gain steps, so each gain step should increase the voltage output by 1.014 volts per step.

$$\text{Gain\_Step} = \frac{\text{Actual Span}}{\text{Number of Steps}} = \frac{7.1 \text{ volts}}{7} = 1.014 \text{ volts}$$

With an input voltage level of 1.00 volts and a gain setting of 1, the DUT generated an output level of 1.05 volts. Changing the gain setting should produce equally spaced increments of 1.014 volts each, with a final value of 8.15 volts. Comparing the output level for each gain setting with the calculated level tests the device linearity.

Table 2.4

Linearity Tests Measure Each Step

Gain Step	Calculated Value	Measured Value	Error %
1	1.05	1.05	0.0%
2	2.064	2.107	+2.1%
3	3.078	3.031	−1.5%
4	4.092	4.051	−1.0%
5	5.106	5.116	+0.2%
6	6.12	6.193	+1.2%
7	7.134	7.060	−1.0%
8	8.15	8.150	0.0%

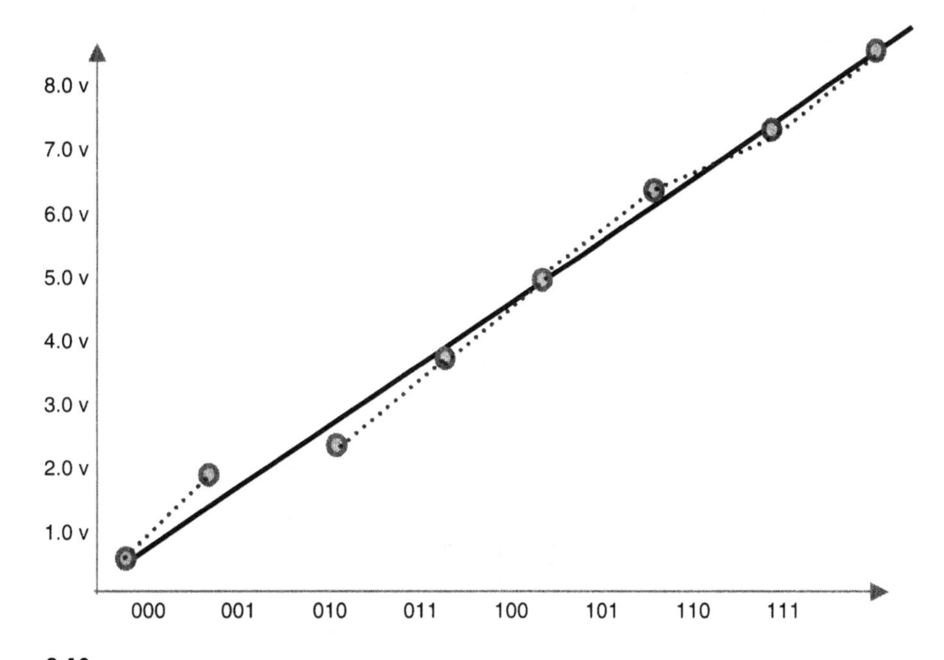

Figure 2.13

Plotting Actual versus Ideal Linearity Response

2.6 Time Domain Tests

Like the DC tests, time domain tests begin with the device specification. In the example case of a programmable gain amplifier, several dynamic characteristics are specified. The purpose of the dynamic specifications and tests is to ensure adequate device signal performance.

Example DUT Time Domain Specifications

Slew Rate = 10 μs per volt
Settling Time = 5 μs
Frequency Response = 100 Hz to 10 kHz +/–4 dB

2.6.1 Slew Rate and Settling Time

The time domain specifications for the example device include slew rate and settling time. Slew rate describes the slope of a voltage change across time. The DUT is driven with a fast edge pulse, and the output is captured and analyzed. The slew rate is found as the slope of the transition between the rated output extremes. Sometimes the positive and negative swings will have different slew rates, in which case both positive and negative slew rates are tested.

Settling time measures the time elapsed from the application of a step input to when the amplifier output has settled to within a specified error band of the final value. Settling time includes the time needed for the DUT to slew from the initial value, recover from any overload, and settle to within a specified range.

Because the test procedures for slew rate and settling time use an identical set of test conditions, the two tests can be efficiently grouped together. In both cases, the device is powered up and programmed for a specific gain value. A square wave signal is applied to the amplifier input, and the amplifier output signal is captured and analyzed.

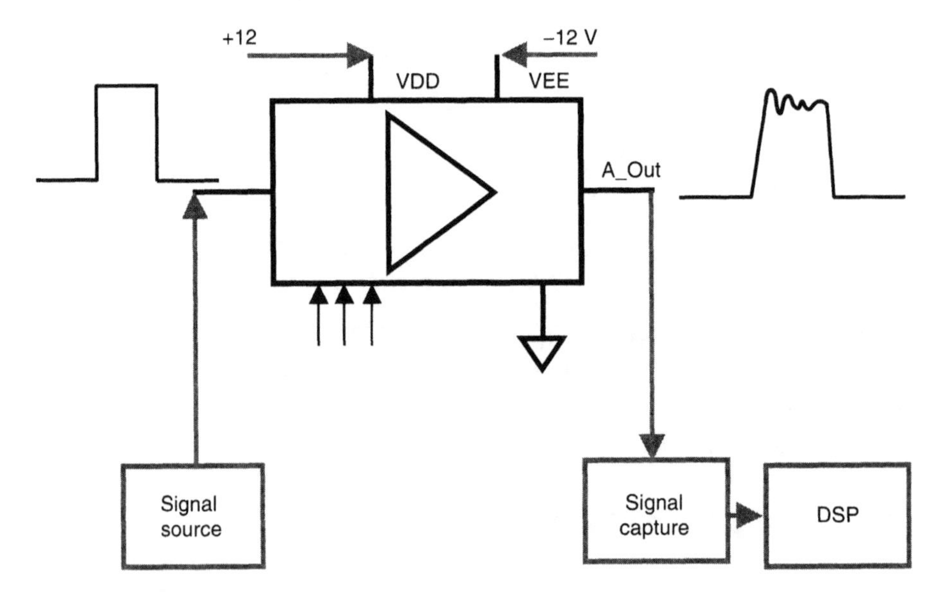

Figure 2.14

Slew Rate and Settling Time Test Setup

The captured signal from the device output is analyzed by the DSP unit. The DSP calculates the slew rate by measuring the period between two thresholds of the signal slope. Settling time is calculated from the captured signal information as the time between the beginning of the slope, and when the output state is within the specified range of the new output level.

$$\frac{\Delta V}{\Delta T}$$

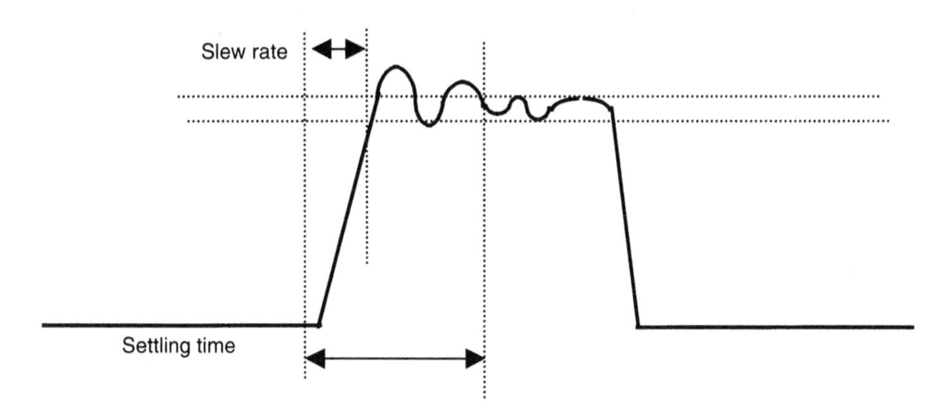

Figure 2.15

Slew Rate and Settling Time Measurements

2.6.2 Frequency Response Tests

Mixed signal devices may be specified to operate over a range of signal frequencies. A frequency response test measures how the device responds to different signal frequencies across a specified range. Frequency response can be measured in either time domain or frequency domain.

Test Procedure

The device is powered up and programmed for a specific gain value. Three separate measurements are performed, using three different input signal frequencies. The first measurement applies a 1000 Hz sine wave, with a 1.0 volt peak-to-peak amplitude. The output signal amplitude is measured at 1.05 volts. The second measurement applies a 5000 Hz sine wave, also at 1.0 volts peak-to-peak. With a 5000 Hz input signal, the output signal amplitude is measured at 0.95 volts. The third measurement applies a 10 kHz signal at 1.0 volts peak-to-peak amplitude, which results in an output amplitude of 0.73 volts.

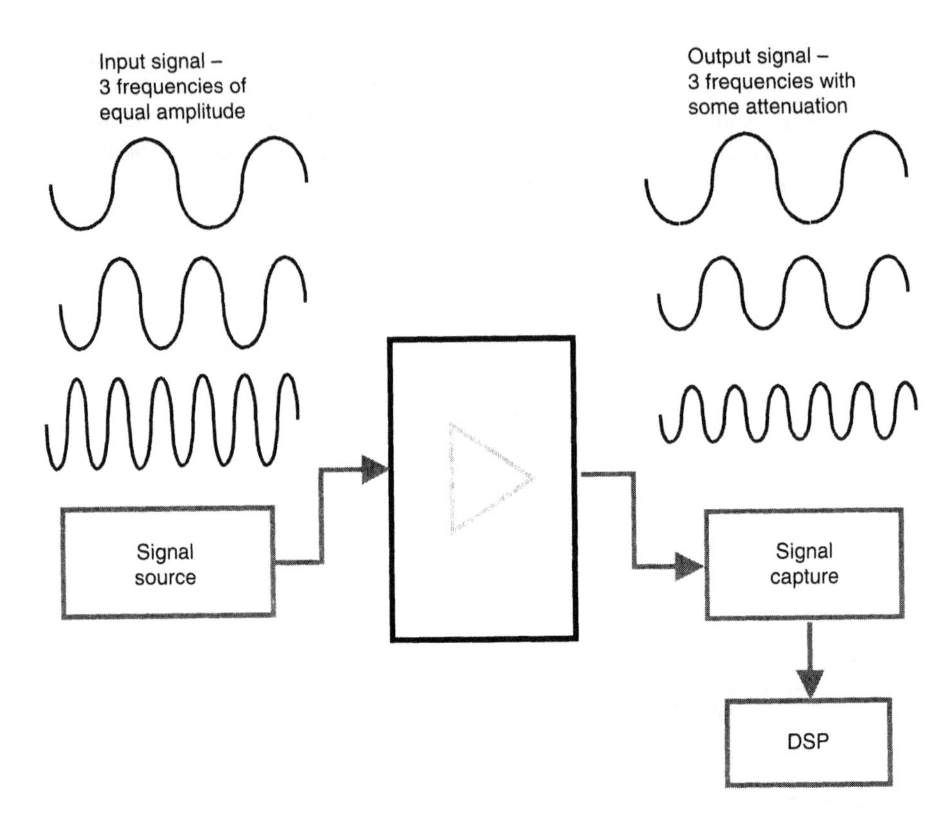

Figure 2.16

Frequency Response Measurement—Time Domain

What's the dB Ratio?

Referenced to the 1000-Hz output signal amplitude, the 5-kHz signal is down by –0.87 dB.

$$\text{dB} = -20 \times \log\left(\frac{1\,\text{kHz Reference}}{5\,\text{kHz Signal}}\right) = -20 \times \log\left(\frac{1.05\ v}{0.95\ v}\right) = -0.87\,\text{dB}$$

For the 10-kHz signal, the attenuation is –3.15 dB.

$$\text{dB} = -20 \times \log\left(\frac{1\,\text{kHz Reference}}{10\,\text{kHz Signal}}\right) = -20 \times \log\left(\frac{1.05\ v}{0.73\ v}\right) = -3.15\,\text{dB}$$

By plotting the results of the three measurements, we can display a graph of amplitude across frequency.

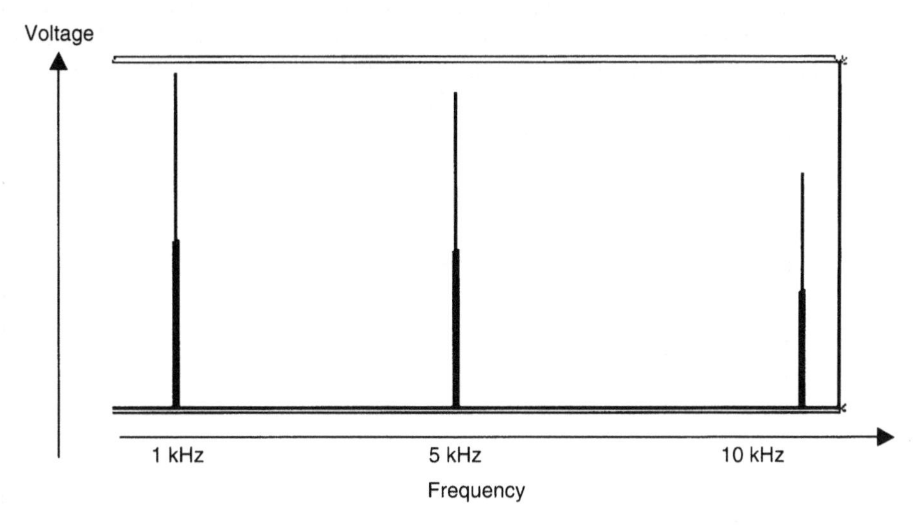

Figure 2.17

Frequency Response Plot

2.7 Frequency Domain Tests

A faster way of measuring frequency response is to use one input signal that is composed of several frequencies, a signal known as multi-tone.

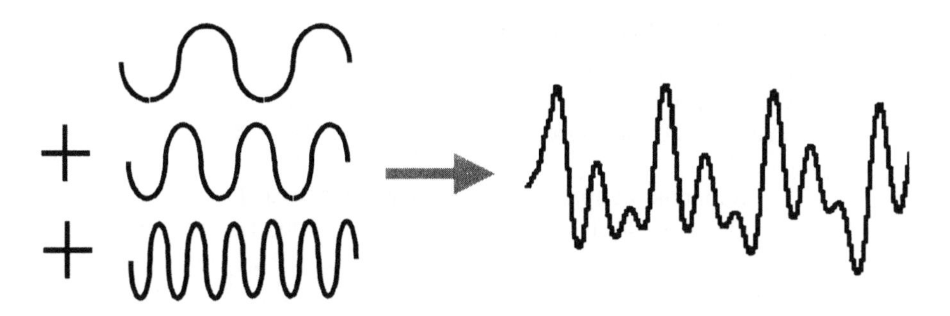

Figure 2.18

Generating a Multi-tone

2.7.1 Multi-Tone Signal

By applying a multi-tone signal, the device response to each frequency component can be evaluated by processing the device output in the frequency domain. Multi-tone testing uses a DSP process that deconstructs the device output signal into a data set of amplitude values at discrete frequencies. The result of the DSP algorithm, known as the Fourier Transform, allows relative measurements of distinct signal frequency components. Using the DSP to generate and process signal data as discrete frequencies is called frequency domain analysis.

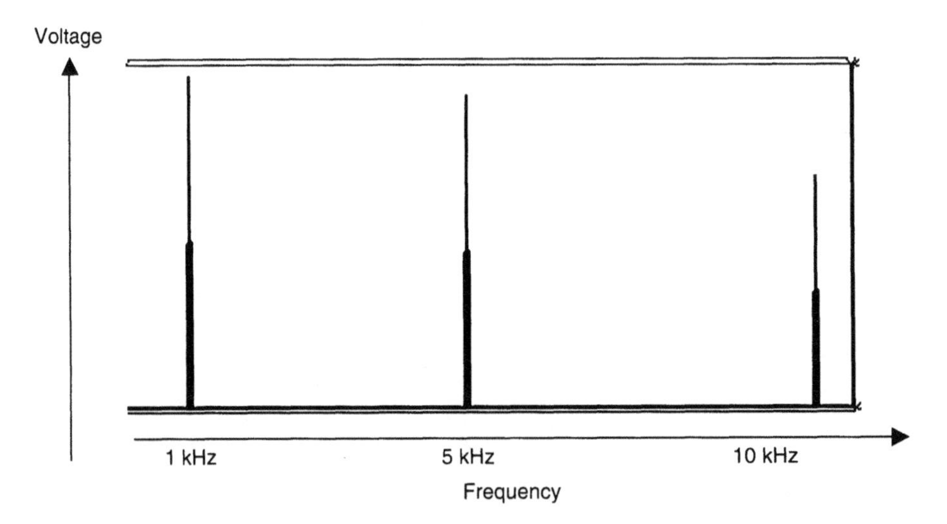

Figure 2.19

Frequency Response Measurement—Frequency Domain

2.7.2 Noise and Distortion

The DUT introduces errors into the signal, which can be analyzed as distortion and noise. The specifications for the example device describe the output signal purity as a maximum allowable error ratio, referenced to the primary signal.

Harmonic Distortion	<5% at 1000 Hz at 1 volt
Signal to Noise	–60 dB with 1000 Hz reference at 1 volt

Distortion is an error in the signal shape, and occurs for each signal cycle. Distortion errors are therefore multiples of the fundamental frequency, and are called harmonics. Noise is random error that is not related to the shape or period of the original signal. Noise is usually defined as spurious signal energy in the output signal of the device, which occurs at non-harmonic intervals of the original signal frequency.

The stimulus data for both distortion tests and noise tests is a single-tone sine wave, known as the fundamental frequency. Assuming that the input signal is relatively pure, any difference between the input signal and the output signal is error introduced by the device.

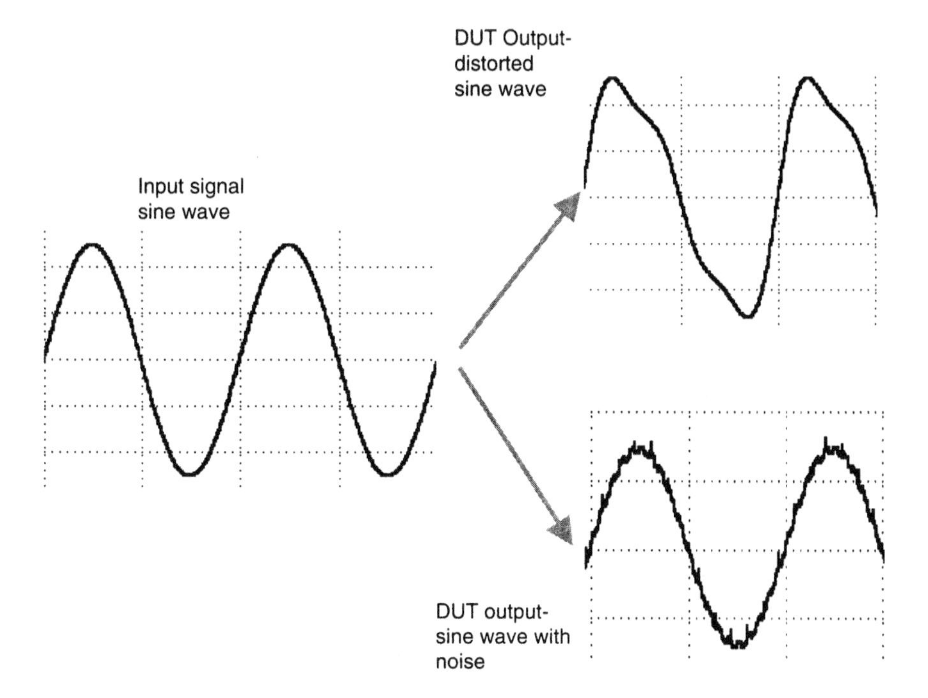

Figure 2.20

Distortion and Noise

2.7.3 Testing for Distortion and Noise

The test process for distortion and noise testing applies a pure sine wave to the DUT. The output of the DUT is captured and processed with a Fourier Transform. The Fourier Transform is a mathematical process that organizes the signal information according to variations of amplitude across frequency. (More on this later!)

By evaluating the frequency domain data, the amplitude of the original signal frequency can be compared with the amplitude of the signal distortion, which occurs at integer multiples of the original frequency. Signal information that is not the original signal frequency and not an integer multiple of the original signal frequency is identified as noise.

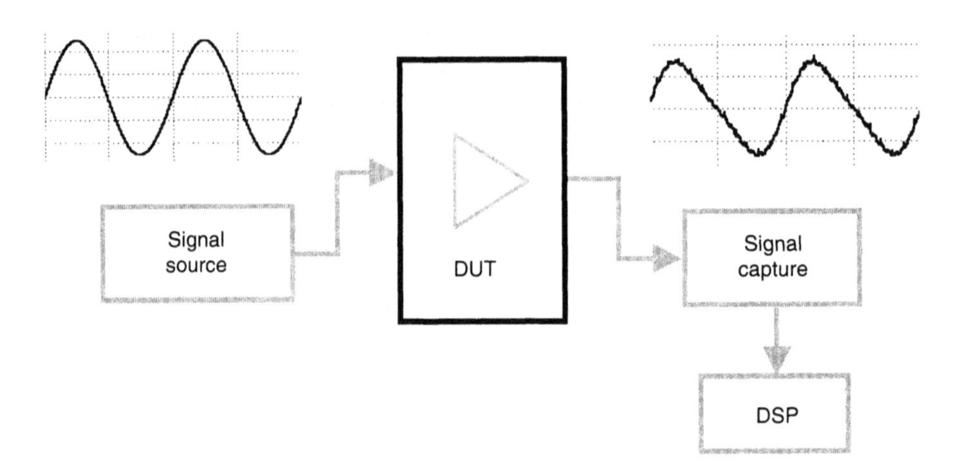

Figure 2.21

Measuring Distortion and Noise with the DSP

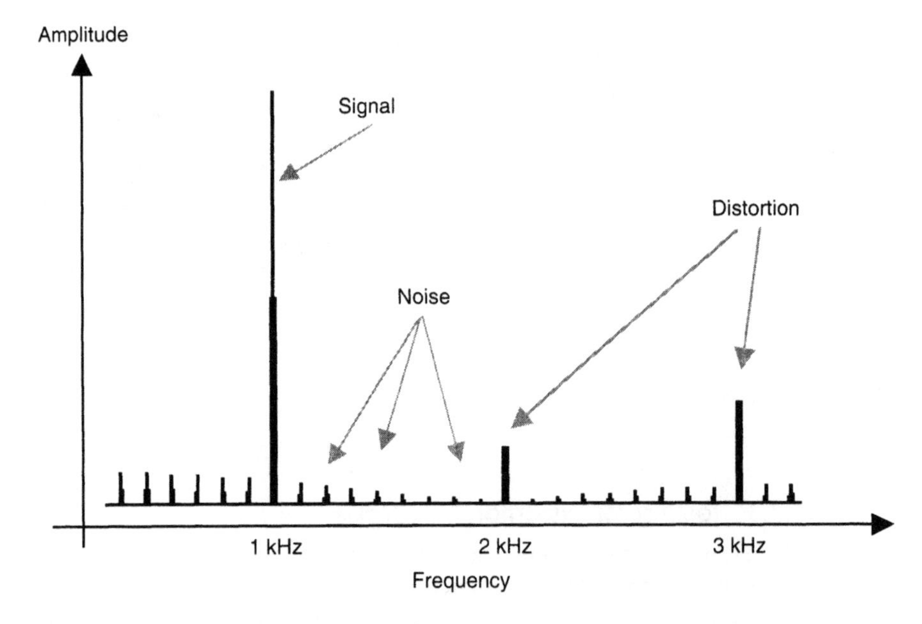

Figure 2.22

Frequency Domain Plot of Signal with Noise and Distortion

Chapter Review Questions

1. List five mixed signal test measurements.

2. Define noise and distortion.

3. Calculate the voltage dB ratio of 13 mV and 1.72 volts.

CHAPTER 3

SIGNAL GENERATION

There still remain three studies suitable for free man. Arithmetic is one of them.

—Plato (ca 429–347 BC)

3.1 Introduction

I've read a few books about sampling theory and signal generation, and most of them give me a headache. Rather than put you through the same kind of pain and agony, this chapter will distill that high-browed theory down to a few very powerful equations. (Maybe distilling is in my genes—let's just say that my grandfather was an entrepreneur doing the Prohibition.) When you master the concepts in this chapter, you'll be well on your way to mastering mixed signal test.

Testing requires the ability to present a stimulus, measure the response, and analyze the results. The mechanism for presenting the stimulus is referred to as **source**. The **source instrumentation** on a mixed signal tester must be able to apply analog data in both analog form and digital form. For example, if the device under test is an analog-to-digital converter, the input data will be in analog form. If the device under test is a digital-to-analog converter, the input data will be in digital form.

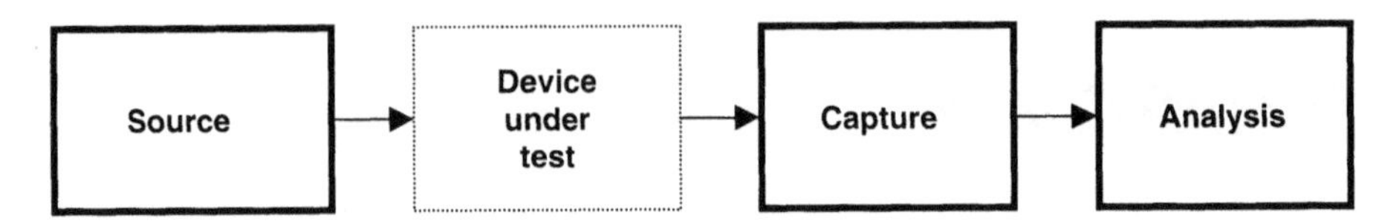

Figure 3.1

The Three Steps of a Mixed Signal Test

A numerical representation of the analog information to be applied to the device is called a **wave shape** or **sample set**. The stimulus wave shape is defined as a series of numeric sample values. A plot of this sequence of numeric samples, the sample set, traces the desired wave shape. The source synthesizes analog signal information from the numeric sample set.

Let's say that the device under test is a digital-to-analog converter. In that case, you'll want to apply the input data as analog information in digital form. If the device under test is an analog-to-digital converter, then the input data is presented as analog information in analog form. Just to keep things interesting, let's not forget that a codec device is basically both an ADC and DAC on one chip. The signal source instrument allows us to present input signals in both analog and digital form at the same time.

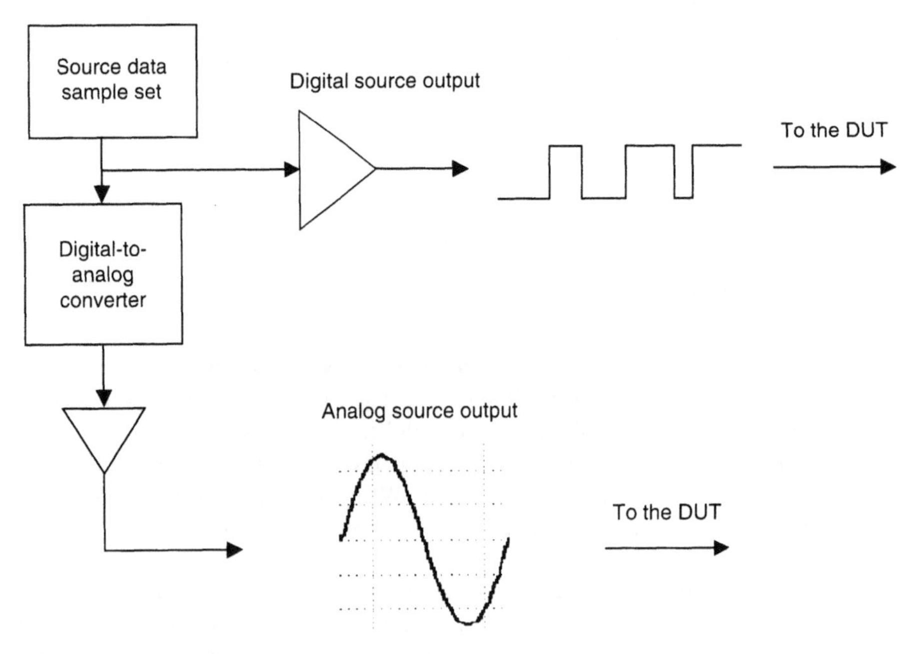

Figure 3.2

Mixed Signal Source—Analog and Digital

3.2 Signal Source Hardware

The signal source converts the signal sample set into a repetitive waveform by continuously incrementing the sequence of samples. If the stimulus waveform is in analog form, the signal source drives a digital-to-analog converter (DAC) to create an analog signal. The output of the DAC is **filtered** to remove digitizing noise, and then amplified to the specified level.

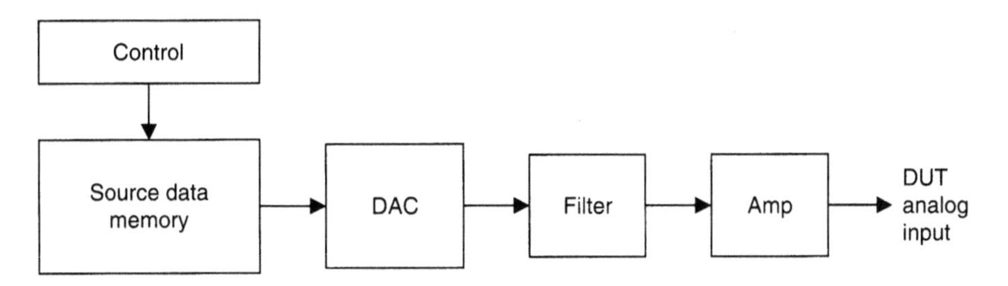

Figure 3.3

Analog Signal Source Block Diagram

If the stimulus waveform is in digital form, the signal source drives the digital pin drivers. The level, format, and timing parameters are the same as for a digital test.

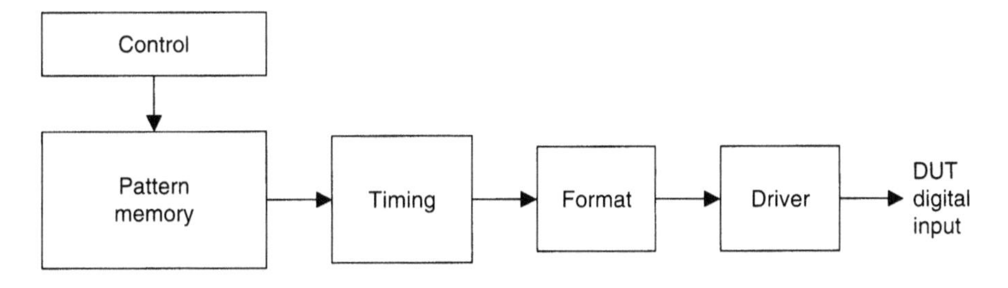

Figure 3.4

Digital Signal Source Block Diagram

3.2.1 Analog and Digital Signal Generation

If the device under test is an ADC, the input will be in analog form and will be generated with the analog instrumentation. If the device under test is a DAC, the input will be in digital form and will be generated with the digital instrumentation.

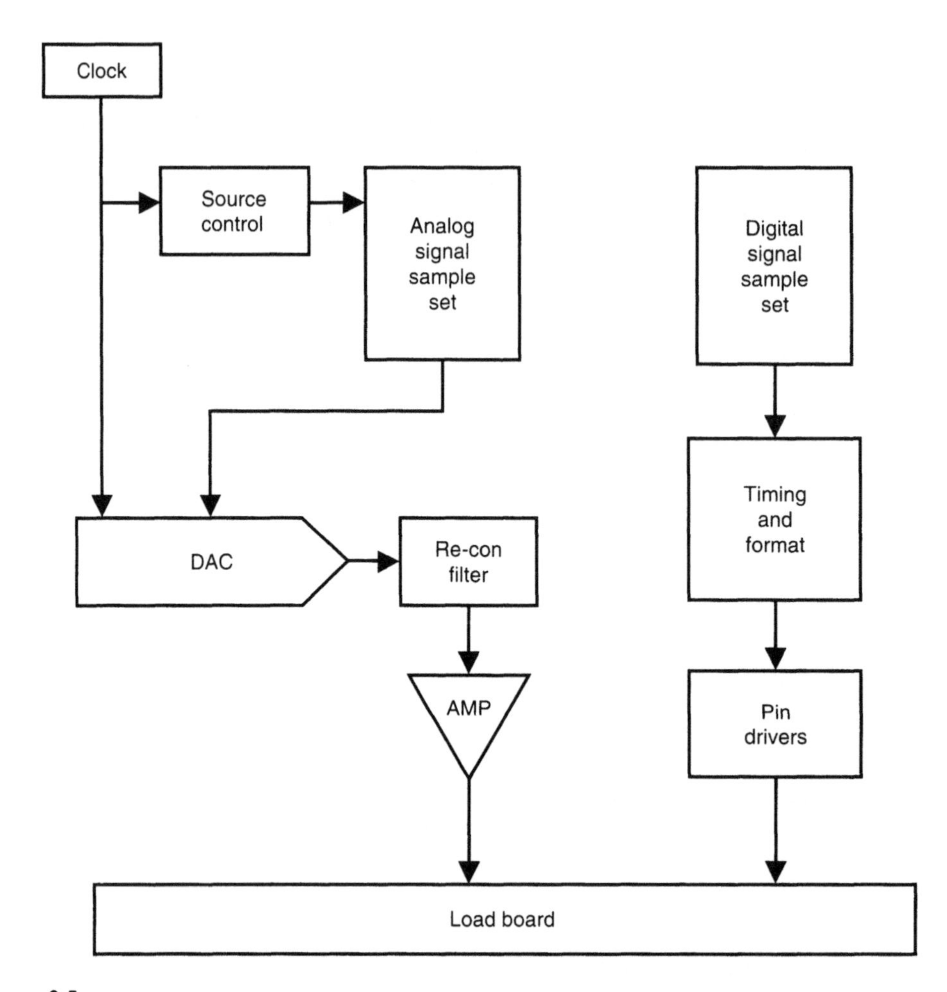

Figure 3.5

Signal Source—Analog and Digital

3.2.2 Controlling the Source

Usually, all of the wave shapes that will be used in a given test program are stored in the source memory when the program is first loaded. In most ATE systems, the source can be started and stopped through either the ATE main control program or through a control field in the pattern.

3.2.3 Test Program Files

A functional test typically uses several files with distinct purposes and structure.

The main **control program** manages the overall test sequence. Typically, the control program manages the overall program flow, the instrument control sequence, and the test result processing. Another file used by the test program contains the "truth table" description of the device logic function. The pattern file is a text description of the digital data sequence to be applied to the device, and the expected digital output data from the device, organized as a set of vectors. The pattern file text is compiled and installed in the tester pattern RAM, and is associated with a unique pattern label.

Control Program File

```
main()
{
initialize_part();
turn_on_power();
continuity_test();
idd_test();
start_wave_source();
run_pattern("CHECK.pat");
}
```

Pattern File

Pattern CHECK.pat

; #	tset	CTRL	vector
#01	ts1		0011LLHH
#02	ts1		1010HHLH
#03	ts1		1110LHLH
#04	ts2	start_src	1110LHLH
#05	ts1		0011LLHH

Main Program control example

```
start wave_source_1;  /* program start command */
```

Pattern control example

```
/* address  command  tset  code  vector    */
```

Table 3.1

Controlling the Source Instrument

3.2.4 Digital Source Circuit Description

The digital signal source instrumentation provides analog information in digital form to the device input. A binary representation of the device functional pattern is stored into test system signal source RAM. The digital signal sequence is programmed as a series of **vectors**. The pattern file description for each vector includes

the binary data to be applied to the DUT inputs
the vector control command
the selected timing set

Example Pattern File Vector Set		
/* command	**timing set**	**digital signal */**
STEP	TS1	01101010
STEP	TS1	10101011
STEP	TS1	01011011
STEP	TS1	10010101

The digital signal source memory is accessed by the sequence controller. The sequence controller looks up the command and timing information for the selected vector and applies the timing information to the formatter. Formatting determines the edge placement timing within the vector cycle of the data presented to the device input pins via the pin driver. The pin driver acts as a high-speed switch that converts the formatted data into voltage levels representing the binary signal data.

3.2.5 Analog Source Circuit Description

The **Clock** is programmed to the correct sample rate and drives the **Sequencer** and the digital-to-analog converter (DAC).

The Sequencer steps through addresses in the **Source Memory**.

The Source Memory provides the sequence of digital samples to the DAC.

The DAC circuit converts the digital samples into analog levels.

The **Reconstruction Filter** smoothes the sequence of discrete analog levels into a continuous analog signal.

The **Amplifier** adjusts the level of the signal required by the DUT.

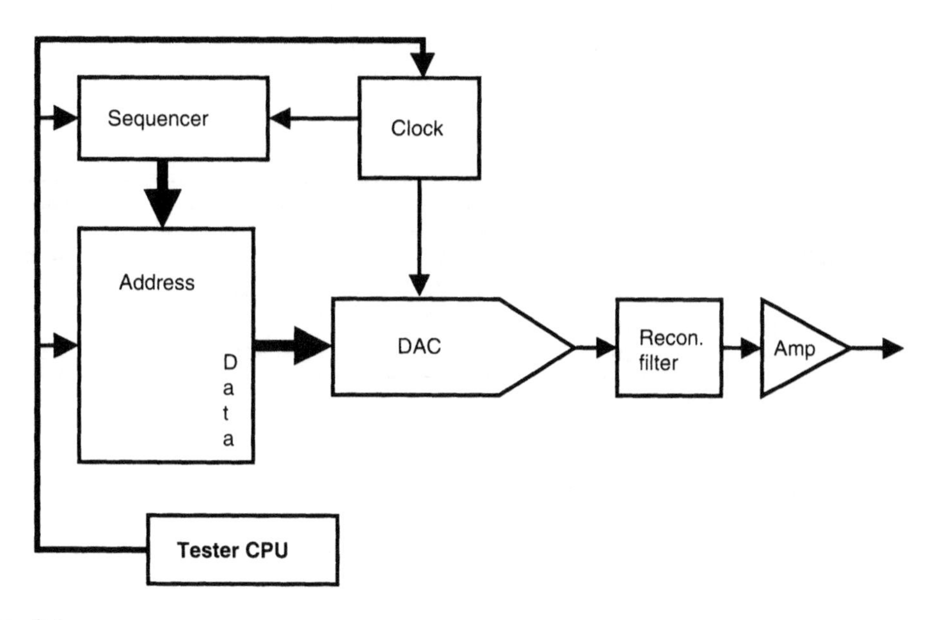

Figure 3.6

Analog Signal Source

The purpose of the analog signal source is to provide an analog signal of a specific shape, frequency, and amplitude to the DUT. To program the wave shape, a sequence of numeric data points is stored in the source memory. Typical ATE signal source units have a source memory from approximately 256K to 1 MG deep by 20 bits wides. The source memory can store many unique sequences—for example, one sequence for a sine wave, one for a triangle wave, and one for a saw tooth wave. Usually the source memory is loaded with the required wave shape sequences when the test program is first initialized.

The analog signal source sequencer is controlled by the ATE system software to loop on the selected sample set in the source memory. Many ATE systems feature sequencers that include sub-routine jumps and similar "state machine" functions. The rate at which the sequencer steps through the source memory is controlled by the source clock. Data from the source memory is loaded in the DAC, which is also controlled by the source clock. The conversion speed and amplitude resolution of the DAC characterizes the target application of the analog signal source. Table 3.2 summarizes the approximate speed and resolution of typical ATE source systems.

The reconstruction filter is so named because it interpolates between data points to produce a continuous signal. The low pass filter rejects the higher frequency components of the sample steps, smoothing the sequence of discrete analog levels into a continuous analog signal. ATE systems typically feature a bank of filters with different cut-off frequencies. The amplifier provides a programmable gain function. The level of the signal is adjusted to the amplitude required by the DUT.

Table 3.2

Source Instrumentation Options

# of Bits	Resolution @ 1V	Conversion Rate	Application
8	+/– 4 mV	200 MHz	Very high speed
12	+/– 24 4uV	50 MHz	High speed
16	+/– 15 uV	1 MHz	High accuracy
20	+/– 1.0 uV	200 KHz	Very high accuracy

3.3 Application Example

Let's say you program the source to generate a 1-volt peak 1000-Hz sine wave. Later, you decide you need a 2000-Hz 1-volt sine wave. Does that mean you must reprogram the signal data set? What if you needed a 2-volt sine wave? Would you have to reprogram the signal data set in that case?

Answer: You would not need to reprogram the data set in either case. You can reprogram the clock rate to generate a different frequency, and reprogram the amplifier range to produce a different amplitude level.

3.4 Signal Data Sets

A periodic sample set is a numerical replica of the desired wave shape that can be repeated continuously to generate a continuous waveform.

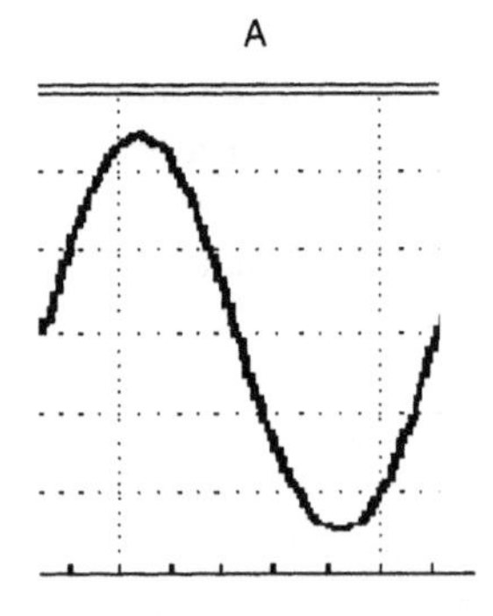

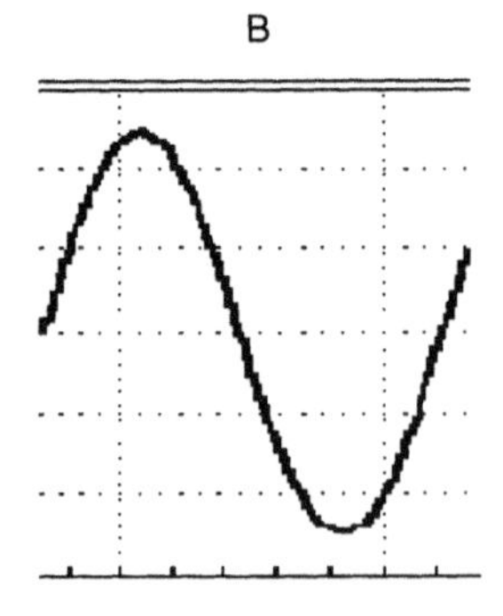

Figure 3.7

Single Cycle and Non-Integer Cycle Sample Sets

3.5 Periodic Sample Sets

Both waveform "A" and waveform "B" are data sets that appear to represent periodic sine waves. However, if the data set is repeated, it becomes clear that the waveform "A" is periodic, and the waveform "B" is non-periodic. Repeated iterations of sample set "B" misrepresent the wave shape and introduce discontinuities. Periodic sample sets require an integer number of waveform cycles.

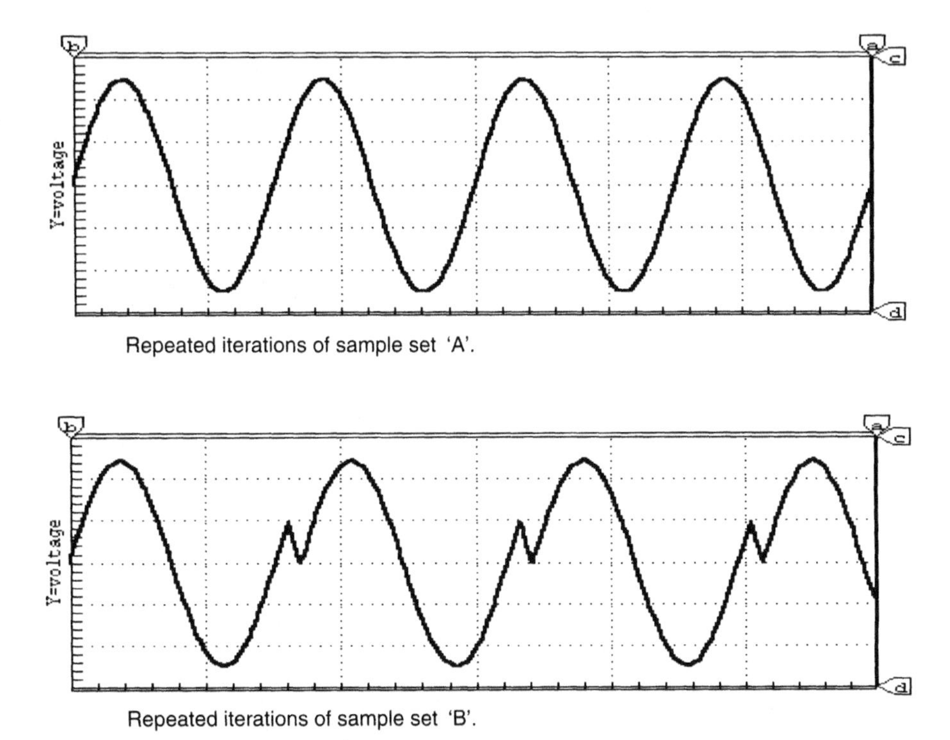

Repeated iterations of sample set 'A'.

Repeated iterations of sample set 'B'.

Figure 3.8

Repeated Sample Sets: Periodic and Non-Periodic

Pretty ugly, huh? What happened to our neat little sine wave? The source instrument will loop on the data set, which effectively stacks the waveform, end-to-end. If the endpoints of the waveform don't line up, then you'll wind up with these hellacious lumps and bumps in your signal set. The technical term for those little boogers is a discontinuity.

3.5.1 Periodic Data in the Frequency Domain

Periodic sample sets are very important in terms of signal generation, and are also critical for proper signal capture and analysis. As we will see in Chapter 5, non-periodic sample sets can significantly corrupt the analysis process. If you thought discontinuities were ugly in the time domain, it's even worse in the frequency domain. It's a really bad deal, because it messes up the frequency domain data set to the point where there's not much useful information.

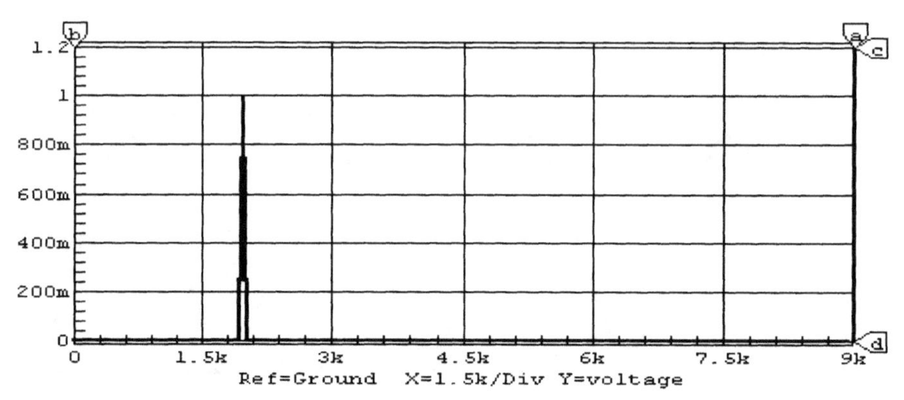

Periodic signal set in the frequency domain

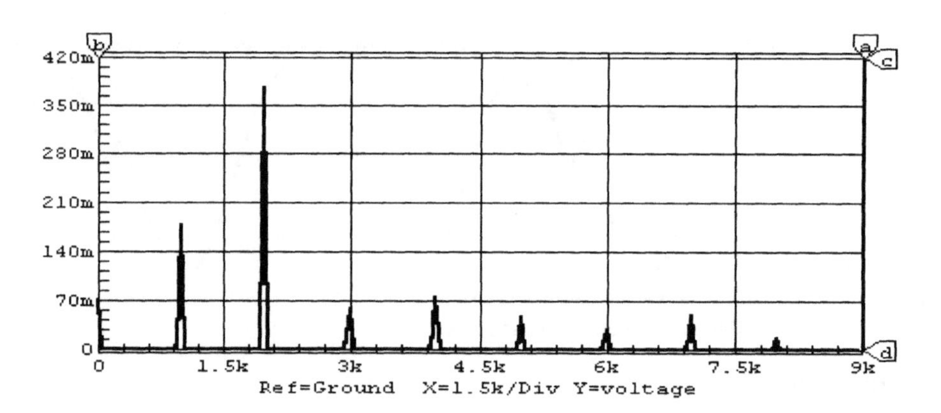

Figure 3.9

Frequency Domain Response to Non-Periodic Data Sets

3.6 Creating the Signal Data Set

Creating a digital model of the wave shape consists of defining a numerical replica of the desired signal. An array of floating point values is used as the software model of the desired waveform. A typical program process uses a repetitive loop to step through each value in the array. Let's take a look at a C code example.

3.6.1 Data as an Array

C code example:

```
float wave_array[16];
int x=0; int sample_size=16;
for (x=0; x<sample_size; x++)
    {
    wave_array[x] =???????? * x; /* wave form equation */
```

0.0	0.3	0.5	0.7	1.0	0.7	0.5	0.3	0.0	−0.3	−0.5	−0.7	−1.0	−0.7	−0.5	−0.3

Signal Array

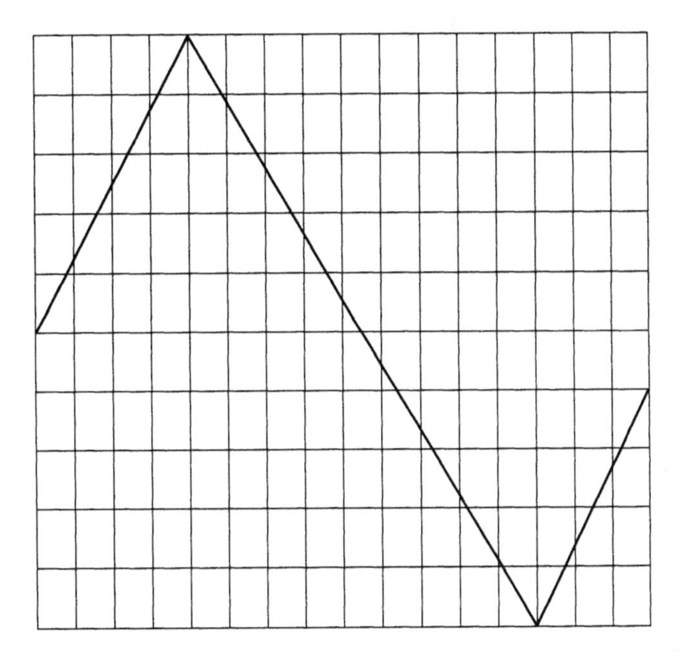

Figure 3.10

Simple Waveform Plot

The array is a numeric model of the signal shape. Notice that the value of the last element of the array, and the last point of the plot, is not the same as the first value. Anticipating that the data represents a periodic set, the last value is one step behind the first point of the next cycle.

Even though the data set appears to be asymmetrical, viewing the data set as a repeated sequence shows that the last data point in the array corresponds to the last data point in the wave shape data set.

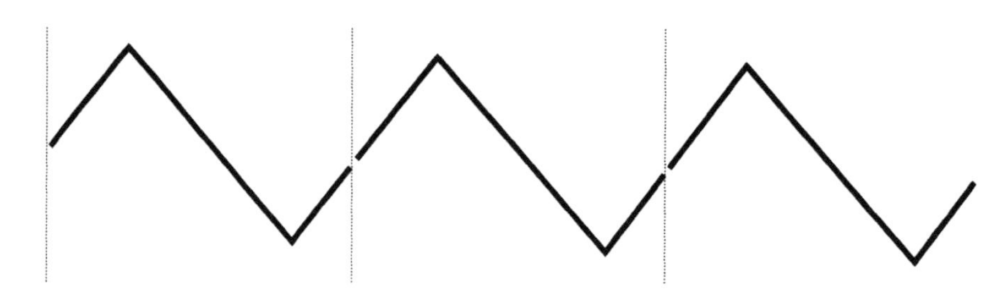

Figure 3.11

Correctly Terminated Sample Set

Because each step through the array represents an increment in the x-axis of the wave shape, a data set that appears to be symmetrical produces unintended results when viewed as a repeated sequence. If the numeric model of the wave shape begins and ends with the same amplitude value, the same amplitude value is repeated twice in succession. The result is a "flat spot" in the wave form plot when repeating the signal data sequence.

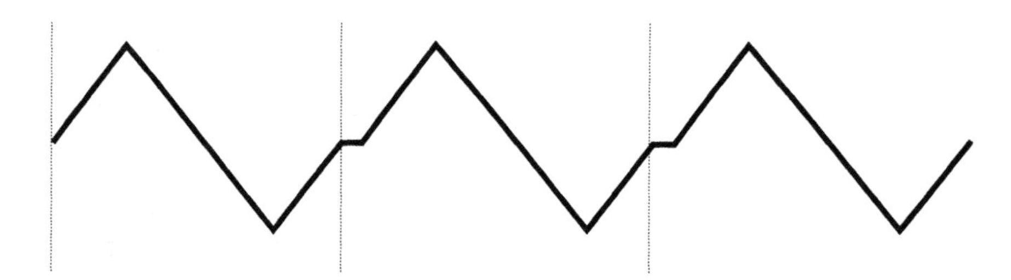

Figure 3.12

Incorrectly Terminated Sample Set

3.6.2 The Sine Wave Equation

Sine waves are extremely useful in mixed signal test applications because all of the signal energy is concentrated into a single frequency. There are no discontinuities from any adjacent points in a sine wave sample set, because a sine wave is essentially a circle with a unidirectional x-axis.

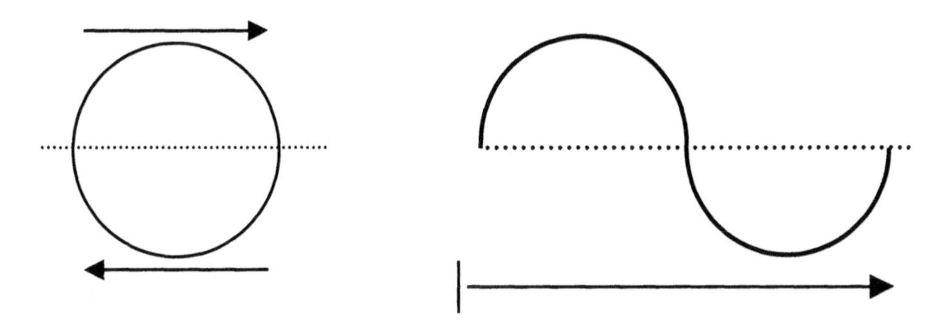

Figure 3.13

Why We Like Sine Waves

The SINE WAVE (Single TONE) Equation goes

```
float new_wave[16];
int x=0; int sample size=16;
float fi=1000.0;
float fs=16000.0;
float PI=3.14159265359;
for (x=0; x<sample_size; x++)
   {
   new_wave[x] = sin(2* PI * x * (fi / fs));
   }
```

What's in the equation?

PI The constant for calculating a circle.
x The loop counter to increment through the array.
fi The frequency of interest, or the signal frequency.
fs The sample frequency, or the clock rate.

3.6.3 The Sample Set Duration

Creating a periodic sample set that produces a periodic signal with the desired wave shape and frequency requires two steps.

1. Determine the formula for representing the wave shape.
   For example, a text-based language formula for a sine wave would be

```
for (x=0; x<sample_size; x++)
       {
       new_array[x] = sin(2*PI*x* (fi/fs));
       }
```

2. Determine the correct value for both the **sample size** and **sample frequency** that will produce a periodic sample set.

In the previous examples, we used a simple case to illustrate the relationships between the sample size, the sample frequency, and the frequency of interest.

sample size = 16
sample frequency = 16000 Hz
frequency of interest = 1000 Hz

The sample period is found by

$$\text{sample period} = \frac{1}{\text{fs}} = \frac{1}{16000\ \text{Hz}} = 62.5\ \mu\text{S}$$

Because the sample set is 16 samples in duration, the duration of the sample set, in time, is equal to $16 \times 62.5\ \mu\text{S} = 1.0$ mS.

What is the period of the frequency of interest?

$$\text{signal period} = \frac{1}{\text{fi}} = \frac{1}{1000\ \text{Hz}} = 1.0\ \text{mS}$$

Wow, they match. The duration of the sample set must be aligned with the signal period.

3.7 DSP's Law

Let's consider the relationship between the sample set duration and the signal period with another example using the following parameters:

fi = 2000 Hz (fi is for frequency of interest. It is the frequency of the signal.)
fs = 16000 Hz (fs is for sample frequency. It is the "step rate" for each data point.)
sample size = 32 (sample size is the total number of samples in the sample set.)

Step One: Calculate the sample frequency period as $\frac{1}{\text{fs}} = \frac{1}{16000\ \text{Hz}} = 62.5\ \mu\text{S}$
Step Two: The duration of the sample window is equal to

Samples × sample period = sample window duration

32 samples × 62.5 μs = 2 ms

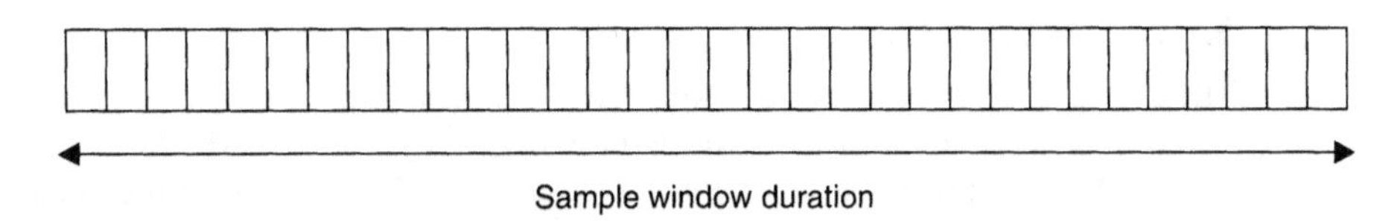

Figure 3.14

The Sample Window—Duration Across Time

Step Three: Calculate the signal frequency period as $\frac{1}{fi} = \frac{1}{2000\ Hz} = 0.5\ mS$

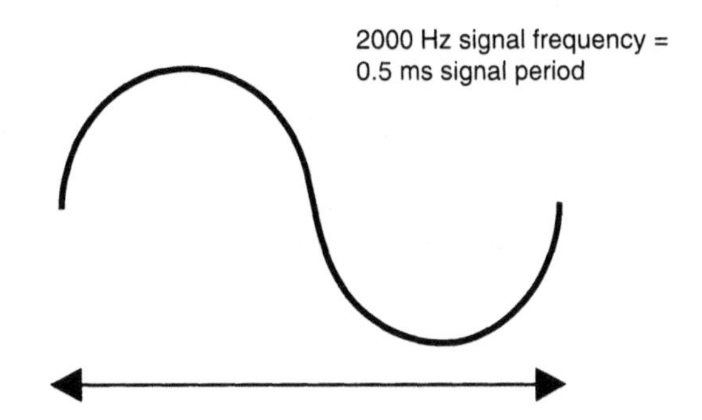

Figure 3.15

The Signal Period

Step Four: Divide the sample window duration by the signal period to determine how many signal cycles will occur within the sample window.

$$\text{signal cycles} = \frac{\text{window period}}{\text{signal period}} = \frac{2.0\ mS}{0.5\ mS} = 4$$

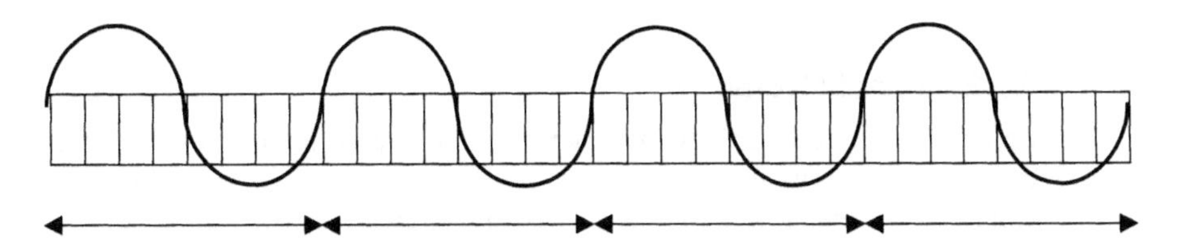

Figure 3.16

Making It All Fit

3.7.1 The Base Frequency (fbase)

The sample set duration can be expressed by calculating a periodic frequency. We will call this periodic frequency **fbase** (base frequency). The OHM's Law equivalent of mixed signal test is the equation for calculating fbase. Because the equation is as simple and as powerful as Ohm's law, we will call the equation DSP's Law.

$$\text{fbase} = \frac{\text{fs}}{\text{samples}}$$

where fs is the sample frequency, and samples is the number of samples in the sample set (or, the number of elements in the data set). If I had to choose one thing in this book that I'd like you to remember ten years from now, this is it.

The fbase expression is simply a way of describing the sample set duration as a periodic frequency. We will see that expressing the sample window as an equivalent frequency has some significant advantages in both time domain and frequency domain. Because the period duration of the sample set must be aligned with the signal period, the fbase periodic frequency must be aligned with the signal frequency (fi).

Let's experiment with the values from the previous example.

fi = 2000 Hz
fs = 16000 Hz
sample size = 32

The base frequency is calculated as

$$\text{fbase} = \frac{\text{fs}}{\text{samples}} = \frac{16000\ \text{Hz}}{32} = 500\ \text{Hz}$$

3.7.2 Signal Cycles

Remember, fbase is a **ratio**. We can also call it a periodic frequency representing the sample set duration. By finding the ratio of the signal frequency to the fbase value, we can determine the number of signal cycles in the sample set.

$$\text{cycles} = \frac{\text{fi}}{\text{fbase}} = \frac{2000\ \text{Hz}}{500\ \text{Hz}} = 4\ \text{cycles}$$

3.7.3 Equation Summary

$$fbase = \frac{fs}{samples}$$

Therefore

$$samples = \left(\frac{fs}{fbase}\right)$$

and

$$fs = fbase \times samples$$

the Cycles Equation is

$$cycles = \frac{fi}{fbase}$$

3.8 Samples per Cycle

There are several effects of waveform sampling that are dependent on the number of samples per cycle. Samples-per-cycle tells us how many points in each cycle of the wave shape. (Remember that you can have more than one cycle of the signal in the sample set!)

Because the number of data points (samples) across time is determined by the sample frequency, calculating sample per cycle is a simple ratio.

$$samples\_per\_cycle = \frac{fs}{fi}$$

$$\downarrow$$

$$fs:fi$$

In some applications, the ratio of fs: fi is used instead of the value.

single cycle example

$sample\_size = 16$

$fi = 1000\ Hz$ $\qquad \frac{fs}{fi} = \frac{16000\ Hz}{1000\ Hz} = \frac{16}{1}$

$fs = 16000\ Hz$

multiple cycles example

sample_size = 320

fi = 1025 Hz

fs = 8000 Hz

$$\longrightarrow \quad \frac{fs}{fi} = \frac{8000\ Hz}{1025\ Hz} = \frac{320}{41}$$

samples per cycle = fs/fi = 8000/1025 = 7.8048

That is, the number of samples per cycle value is 7.8048, but the ratio is 320:41. We will see that the ratio of fs to fi has some useful applications.

3.9 The Golden Ratio

Consider the relationships between the fbase value and the samples per cycle value.

$$fbase = \left(\frac{fs}{samples}\right) = \frac{8000\ Hz}{320} = 25\ Hz$$

$$cycles = \frac{fi}{fbase} = \frac{1025\ Hz}{25\ Hz} = 41\ cycles$$

Even though the number of samples per cycle is a non-integer, the total number of cycles in the entire sample set is the periodic sample set, with a whole number of signal cycles. A common misunderstanding supposes that the ratio of fs/fi must be an integer, but that is not correct. The ratio of fs/fi can be a non-integer as long as the sample set represents an integer number of cycles.

The sample size, sample frequency, frequency of interest, and the number of cycles are all related. As a ratio, we can state that

fs:fi = samples:cycles

Some texts[1] on sampling theory refer to this ratio as

$$\left(\frac{\mathbf{M}}{\mathbf{N}}\right) = \left(\frac{\mathbf{ft}}{\mathbf{fs}}\right)$$

Where M = number of cycles, N = number of samples, ft is the test frequency (what we call fi), and fs is the sample frequency.

[1] A good time for me to tip my hat to Matt Mahoney, author of *DSP Based Testing*.

3.9.1 A War Story

Sounds simple, right? Allow me to tell you the story of Bob and the Bad Cycle Count. Let's say you are the test engineer for a chip company called Bits-R-Us; and that your project device has the very same specifications we've been discussing:

Sample Frequency (fs) = 8000 Hz   Signal Frequency (fi) = 1025 Hz

In the test program, you selected a sample size of 320 samples, which means the test program will generate 41 cycles.

Then Bob, the product engineer, comes along and says, "Kiddo, I think we should have more samples per cycle. I want you to change the sample frequency from eight kilohertz and crank it on up to sixteen kilohertz." You're about ready to tell Bob that your name isn't "Kiddo" when he says, "I'm planning to say some great things about you to your boss when review time comes around." So you figure a few brownie points never hurt. You tell Bob, "Well, I'd be glad to help you out, but it's going to take a lot of work," and go back to your desk.

As the Perl masters will tell you, the three primary qualities of a programmer (and by extension, a test engineer) are sloth, hubris, and impatience. Like all good and lazy programmers, you have designed the program so that the sample frequency is defined with a simple variable statement: fs = 8000.0. You quickly change the declaration to reflect Bob's recommendation, and now the program statement reads: fs = 16000.0. The program re-compiles without any problems and you send it out to the production floor. Pleased with yourself and anticipating all the great stuff that Bob is going to say to your boss, you spend the rest of the day surfing the Web and reading Dilbert.

Your bubble bursts at about two o'clock the following morning, when you get a rather terse phone call from the VP of Operations at Bits-R-Us. He says that production has ground to a halt, and that the test program is generating massive failures. He also points out that this is the last week of the quarter, and that Bits-R-Us is being sued for non-conformance on the very device on which you have butchered the test program.

You get dressed and rush over to the office, swinging by the 7–11 store on the way to grab a cup of coffee. Disheveled and bleary-eyed, you stagger onto the test floor to discover the problem is just as the VP has said. Nothing works. Every part fails the test program, even devices that had tested good before.

What happened? And more importantly, what can you do, since it's now three o'clock in the morning, that will allow you to keep your job and keep your promise to Bob? Well, let's get back to the basics. What does DSP's law tell us about the new parameters?

Not Good News

Recall that the fbase ratio is derived from the sample rate and the sample size. When you changed the sample rate from eight kilohertz to sixteen kilohertz, you also changed the fbase ratio. As you work through the numbers, you realize that Bob is nowhere to be found. He's probably asleep in a warm bed; and *it's all his fault!* Knowing that your befuddled brain cannot be trusted with simple arithmetic at this point, you grab a calculator to run the numbers based on the new fs value.

$$\text{fbase} = \left(\frac{\text{fs}}{\text{samples}}\right) = \frac{16000\ \text{Hz}}{320} = 50\ \text{Hz}$$

The fbase used to be 25 Hz, and now it's 50 Hz. That shouldn't create too much of a problem, right? You're starting to think that maybe you can assign the blame somewhere else, when the second part of DSP's Law presents itself. Read it and weep.

$$\text{cycles} = \left(\frac{\text{fi}}{\text{fbase}}\right) = \frac{1025\ \text{Hz}}{50\ \text{Hz}} = 20.5\ \text{cycles}$$

Uh-oh, there's the problem. The program used to produce a periodic sample set that produces 41 nice clean cycles that repeat over and over again with smooth consistency. Now you have a sample set with a non-integer number of cycles. You've created a hideous wave shape that looks like a sine wave that ran full-speed into a brick wall.

How did that happen? All you did was to double the sample frequency, so what's the big deal? The big deal, as it turns out, is that doubling the sample rate without changing the sample size will cause the window duration to be cut in half. Because the base frequency is twice what it was before, the signal no longer fits. What has to happen to fix the fbase ratio, using the new sample frequency of 16000 Hz? A quick glance at the equation makes it clear.

$$\text{fbase} = \left(\frac{\text{fs}}{\text{samples}}\right)$$

If we double the sample frequency, we must also double the sample size. Inspired with new hope and the adrenaline of outright fear, you edit the program and double the sample size from 320 samples up to 840 samples. The program recompiles, begins to run properly again, and all is well. You mutter something about, "There was a little anomaly from a change that Bob requested," and head out the door.

3.10 Application of DSP's Law

When we create a sample set, usually we will know at least three things:

the shape of the signal
the frequency of the signal
the parameters of the signal source instrumentation

Let's apply some poor man's calculus. We can apply the "Rules of DSP" to create a periodic sample set of the correct shape and frequency. Let's take another look.

$$\text{fbase} = \frac{\text{fs}}{\text{samples}}$$

can be re-arranged as

$$\text{samples} = \left(\frac{\text{fs}}{\text{fbase}}\right)$$

Does the value for samples always have to be an integer? (Hint: Have you ever seen an array with 15.7 elements?) What does that tell us about the relationship between fs and fbase? The ratio of fs and fbase must always be an integer.

Let's also look at our equation for calculating the number of cycles in the sample set,

$$\text{cycles} = \left(\frac{\text{fi}}{\text{fbase}}\right)$$

We know that the number of cycles in the sample set must be an integer to guarantee a periodic sample set. What does that tell us about the relationship between fi and fbase? The ratio of fi and fbase must always be an integer.

We can therefore state that for a periodic sample set, fbase must have an integer relationship with both fs and fi. When creating a sample set, you can often start by choosing an fbase that has an integer relationship with both fs and fi.

Application Example

The ratio of the signal frequency (fi) to the base frequency (fbase) controls the number of cycles in the sample set, and therefore also controls the number of samples per cycle. The ratio of the sample frequency to the frequency of interest (fs/fi) is equal to the number of samples per cycle.

The samples per cycle ratio of fs over fi is equal to the ratio of the sample size over the number of cycles.

$$\text{fs}:\text{fi} = \text{samples}:\text{cycles}$$

Given a required signal frequency of 120 Hz, and a specified sample rate of 1000 Hz, what would be an optimal sample size? Because you know that fbase must be a common denominator for both fs and fi, choose a number that divides into 120 Hz *and* 1000 Hz. The value of 40 Hz should do the trick. Now that you have a usable fbase, you can use DSP's law to determine the number of samples.

$$\text{samples} = \frac{\text{fs}}{\text{fbase}} = \frac{1000\ \text{Hz}}{40\ \text{Hz}} = 25\ \text{samples}$$

$$\text{cycles} = \frac{\text{fi}}{\text{fbase}} = \frac{120\ \text{Hz}}{40\ \text{Hz}} = 3\ \text{cycles}$$

Discussion Question

What if you chose an fbase that was not the largest common denominator? What would be the result of the preceding example if the fbase was 20 Hz?

$$\text{samples} = \frac{\text{fs}}{\text{fbase}} = \frac{1000\ \text{Hz}}{20\ \text{Hz}} = 50\ \text{samples}$$

$$\text{cycles} = \frac{\text{fi}}{\text{fbase}} = \frac{120\ \text{Hz}}{20\ \text{Hz}} = 6\ \text{cycles}$$

In this case, the same three cycles would be repeated twice within the same sample set, for a total of 6 cycles, and 50 samples. If fbase is not the largest common denominator of fs and fi, the signal data set contains redundant information.

Alternative Method

When creating a sample set to generate a periodic signal, it is sometimes useful to calculate the number of cycles and number of samples by processing the ratio of

$$\text{fs}:\text{fi} = \text{samples}:\text{cycles}$$

Given that fi = 120 Hz and fs = 1000 Hz, you can reduce the ratio as follows:

$$fs:fi = 1000:120 =$$
$$100:12 =$$
$$50:6 =$$
$$25:3$$

You can therefore state that the ratio of **samples : cycles** is equal to **25:3**. By reducing the ratio of fs:fi, you know that the sample set will include 3 cycles, and have a total number of 25 samples.

3.11 Sine (X) over X

Now for something completely different. The number of samples per cycle directly affects the quality of the generated signal. Intuitively, we can expect that a higher samples per cycle ratio in the numeric model of the wave shape will produce a higher quality analog signal.

Figure 3.17

Adequate Number of Samples per Cycle

The process of representing a continuous wave shape with a series of discrete steps introduces some signal amplitude degradation. The error of the "curve fit" is a function of the number of samples per cycle, and can be predicted with the following model. The value of x indicates the resolution of the sample per cycle ratio. The sine of x, in radians, divided by x gives a value less than unity that corresponds to the amplitude scalar, "A."

1. $x = 2\Pi\left(\frac{fi}{fs}\right)$

2. $A = \frac{\sin(x)}{x}$

Example

Example: fs = 8000 Hz fi = 1000 Hz

Calculating the value of x. $x = 2\Pi \times \left(\frac{fi}{fs}\right) = 2\Pi \times 125^{e^{-3}} = 0.7853$

Find the sine of x over x, in radians: $A = \frac{\sin(x)}{x} = \frac{\sin(0.7853)}{0.7853}$

$$= \frac{.707}{0.7853} = 0.9003$$

The amplitude reduction in this case would be almost 10%. ATE systems often use calibration routines to correct for the sin (x)/x effects.

3.12 Source Filters

The reconstruction filter of the analog source hardware must be considered when choosing the optimal number of samples per cycle.

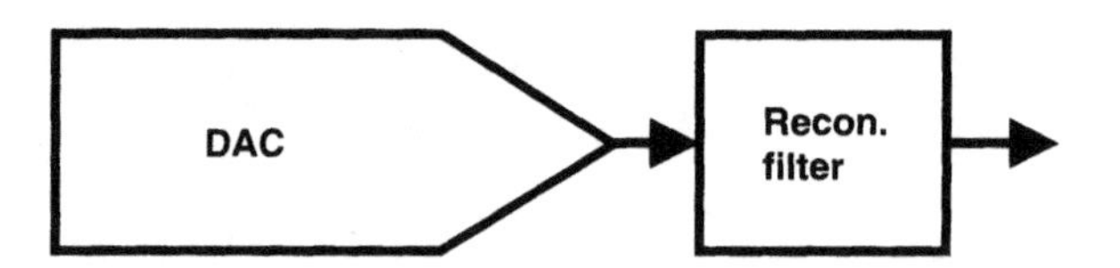

Figure 3.18
The Reconstruction Filter

The purpose of the reconstruction filter is to remove the effects of the sample clock, and the DAC step rate. Typical reconstruction filters are designed to attenuate the clock frequency by 24 dB for every doubling of the pass band. Selecting the 0–1000 Hz filter will pass all frequencies within that range. Frequency components above the pass band would be attenuated as follows:

Source Clock Frequency	Attenuation
2 kHz	–24 dB
4 kHz	–48 dB
8 kHz	–72 dB
16 kHz	–96 dB

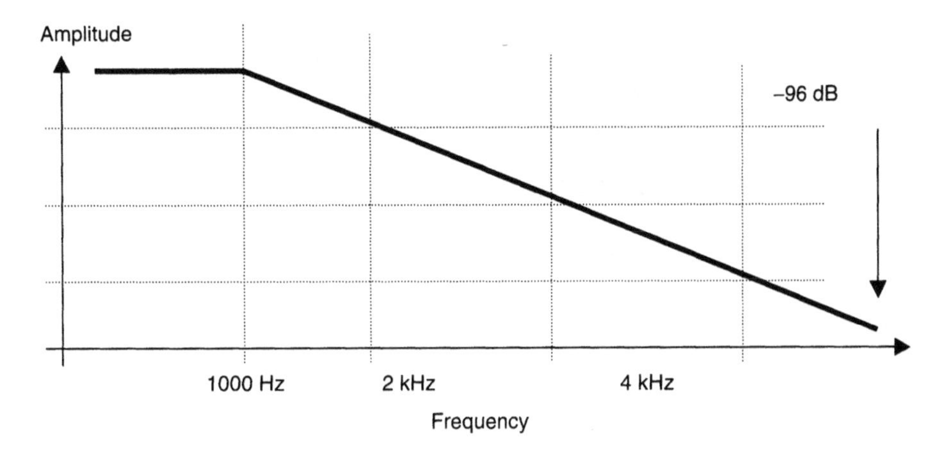

Figure 3.19

Optimal Smoothing Requires a Higher Sampling Rate

3.13 Source Filter Optimization

One of the effects of an inadequate number of samples per cycle is underutilization of the reconstruction filter. A 24 dB per octave reconstruction filter with a 1000 Hz pass band will attenuate a 4000-Hz clock by only 48 dB. The signal generator hardware output signal will contain distortion as a result of the unfiltered clock component. Check out the lumps and bumps in the waveform in Fig. 3.20.

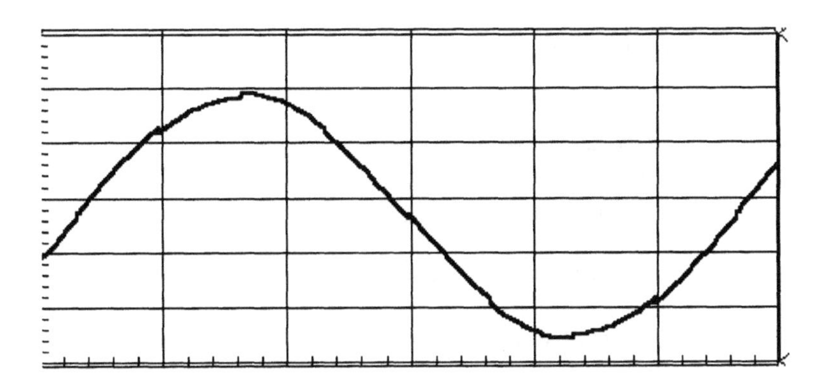

Figure 3.20

Inadequate Filtering Produces Distorted Signals

To derive the best performance from the reconstruction filter in this example, the sample frequency should be 16 times the signal frequency. Actual results will be dependent on the source hardware filter characteristics.

Chapter Review Questions

1. What is the equation to calculate the signal source base frequency (fbase)?

2. You have an application where the Device Under Test (DUT) requires an input of a 1-volt peak triangle wave at a frequency (fi) of 1300 Hz. The test specification says that the waveform must be generated with a sample frequency (fs) of 16300 Hz. Calculate an fbase and a sample size that will generate a periodic sample set for those parameters.

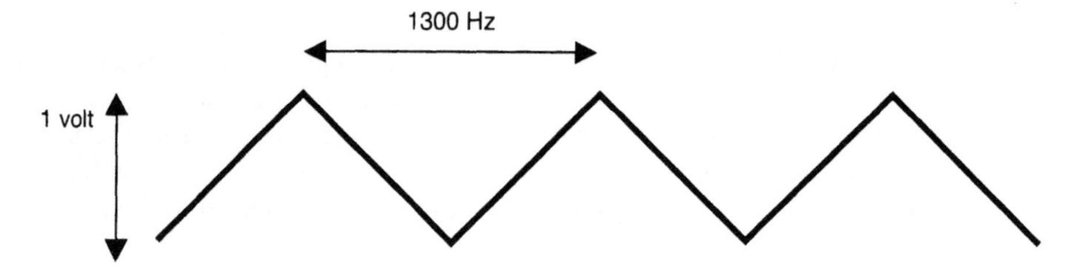

Figure 3.21

A 1300-Hz, 1-Volt Triangle Wave

What is the sample size? __________

What is the fbase? __________

What is the number of cycles in the sample set? __________

4. Aren't you glad that you bought this book?

CHAPTER 4

SIGNAL CAPTURE

Each problem that I solved became a rule which served afterwards to solve other problems.
—Rene Descartes, *Discours de la Methode*

4.1 Introduction

In the last chapter, we looked at the fundamental concepts for generating signal information that will be presented to the DUT input. In this chapter, you'll learn how to use those same concepts and equations to capture the DUT signal output. A mixed signal device may generate analog information in either analog or digital form, which in turn is captured by the ATE systems' signal capture instruments. The signal capture instruments store a numerical replica of the analog information from the DUT in the test systems' capture memory.

If the stimulus waveform is in analog form, the signal capture uses an analog-to-digital converter to create a numeric sample set that represents the original signal data.

The input of the ADC is filtered to remove spurious high-frequency noise, and amplified to the specified range. If the stimulus waveform is in digital form, the signal capture receives the data from the digital pin receivers. The level and timing parameters are the same as for a digital test. Even if the data from the DUT is in digital form, it represents analog information, not a logic function.

The contents of the signal capture memory represent a digitized analog signal, *not digital logic states*. Instead of *comparing* the captured digitized signal with a pattern, the signal data is *analyzed* by a digital signal processor (DSP). The DSP analyzes the signal data to extract analog signal information, such as peak, RMS, signal-to-noise, and harmonic distortion.

In most ATE systems, the signal capture can be started and stopped through either the program or micro-code in the pattern—just like the source.

```
start wave_capture_1; /* program start command */
-OR-
            /* from within the pattern   */
            /* cmd          tset        code              vector          */
               INC          ts2         strt_capt_1       000011 LLLHHHL
```

4.2 Digital Signal Capture Hardware

During a digital capture, the output signals from the DUT are compared against the programmed VOL and VOH levels of the test system's per-pin receiver circuit. The pin receiver outputs indicate if the device signal was a valid logic high (above VOH), a valid logic low (below VOL), or an in-between state (Hi-Z).

For a logic test, timing information selected by the sequencer is sent to the format and timing section, which generates a strobe to the compare circuit. The strobe triggers the compare circuit to compare the DUT logic state with the expected data state described in the pattern vector. In a mixed signal test system, the strobe edge latches the DUT output logic state into the capture memory.

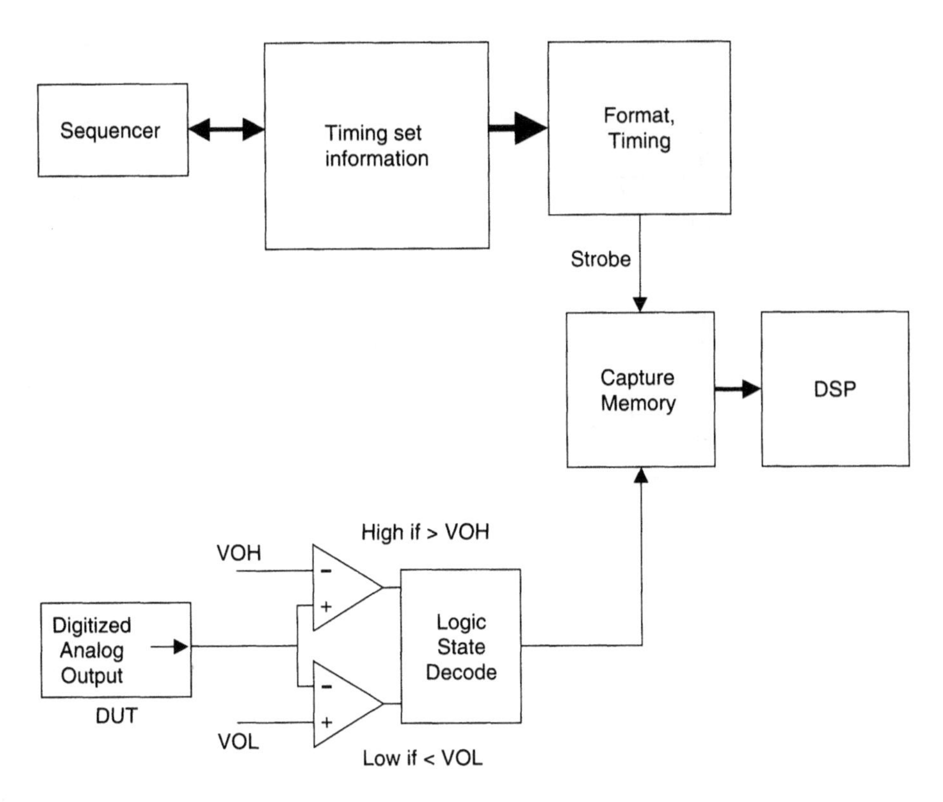

Figure 4.1

Digital Signal Capture Block Diagram

Example: If the device under test is an ADC, the output will be in digital form, and will be captured with the digital instrumentation.

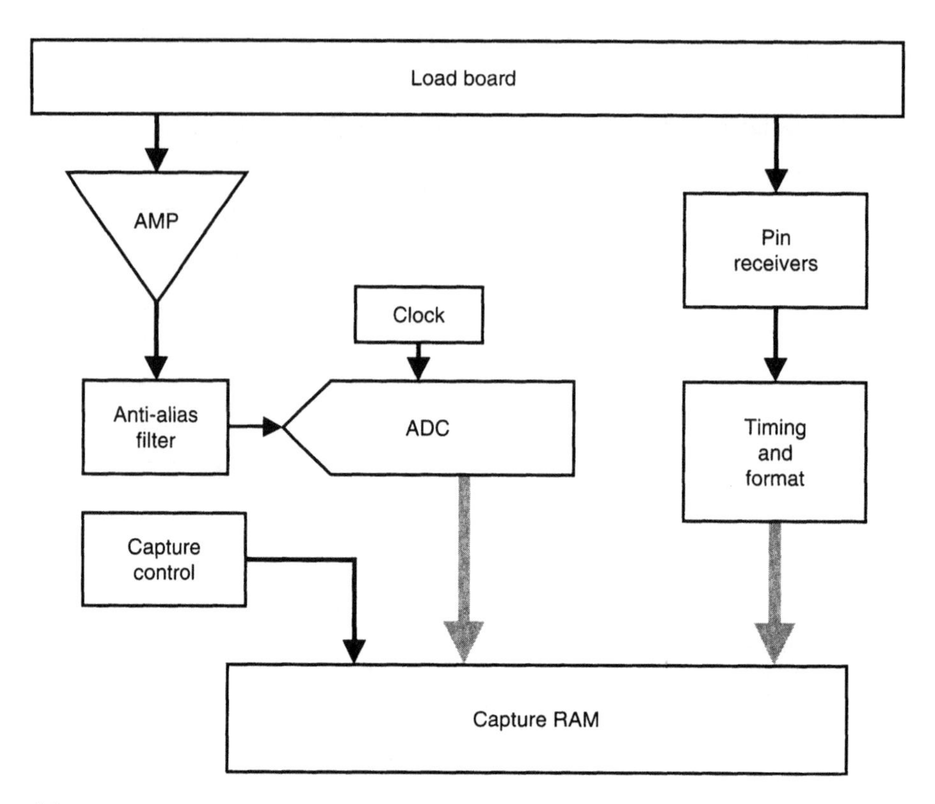

Figure 4.2

Mixed Signal Capture Block Diagram

4.3 Analog Signal Capture Hardware

The purpose of the analog signal capture is to digitize and store the analog signal generated by the DUT for analysis by the DSP unit. The amplifier provides a ranging function. The programmable gain amplifier matches the amplitude of the signal with the fixed range of the ADC. The anti-alias filter rejects spurious high-frequency information before it is digitized. Unwanted high-frequency components can significantly degrade the quality of the signal analysis. Signal information that has been filtered and amplified is digitized by ADC, controlled by the capture clock. The conversion speed and amplitude resolution of the ADC characterize the target application of the analog signal capture. The analog signal capture sequencer is controlled by the ATE system software to store a sequential sample set in the capture memory. The rate at which the sequencer loads data into the capture memory is also controlled by the capture clock. To store the DUT signal, a sequence of numeric data points from the ADC is stored in the capture memory. Typical ATE signal capture units have a capture memory from 256 K to 1 MB deep by 20 bits wides.

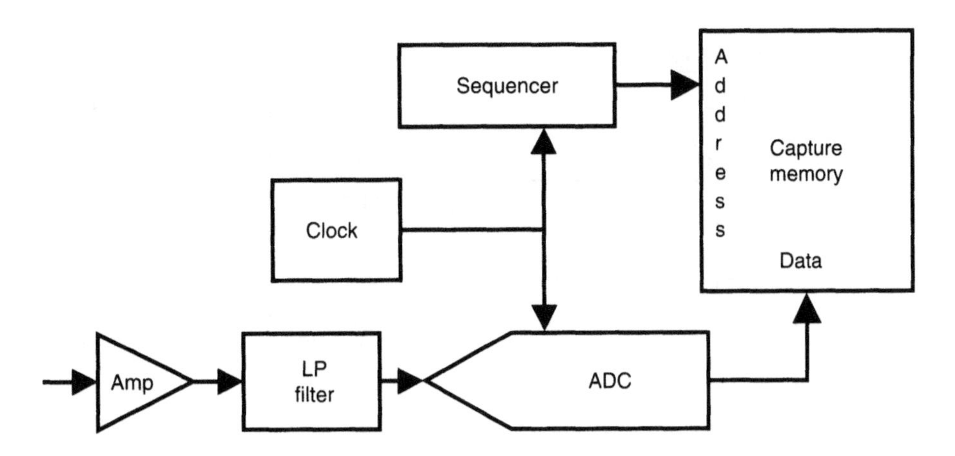

Figure 4.3

Analog Capture/digitizer

Table 4.1

Digitizer Instrumentation Performance Table

# of Bits	Resolution @ 1V	Conversion Rate	Application
12	+/– 244 μV	50 MHz	High speed
16	+/– 15 μV	1.2 MHz	High accuracy
20	+/– 1.0 μV	100 KHz	Very high accuracy

4.4 The Digitizing Process

Digitizing an analog signal represents a continuous signal with a series of discrete numeric values. Each value is called a sample. To digitize means to sample and quantify. This implies that not all data points on the continuous signal are captured. Digitizing attempts to represent a continuous signal with a series of discrete steps. The process of converting from analog to digital has several inherent constraints.

4.4.1 Quantizing Error

Quantizing error is non-random noise of fixed value that is caused by the uncertainty of the digitizing process. A simple two-bit ADC illustrates quantizing error. If the sampled analog value is between two digital levels, the uncertainty is expressed as quantizing error. What happens when the input signal does not correspond exactly to one of the threshold combinations? If the input signal level was at 2.5 volts, for example, the ADC would generate either a 11 or 10 code; and both cases would be in error. (And no, we don't get to have a 11 and a half.) We're stuck between bits.

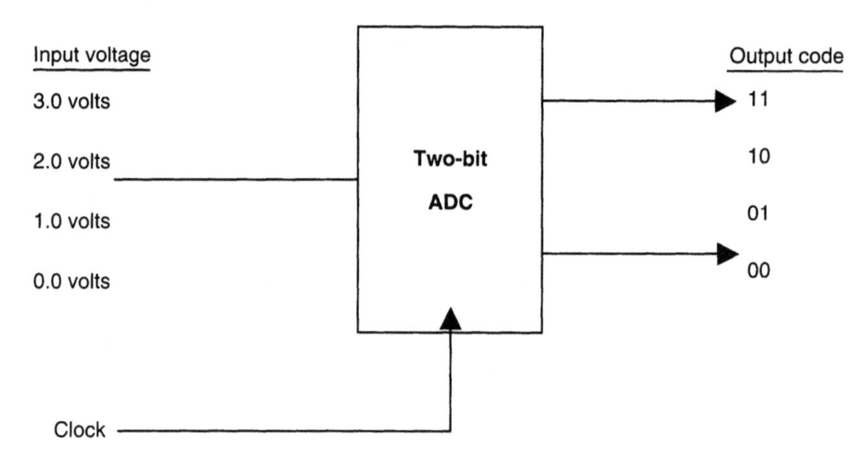

Figure 4.4

Quantizing Error Example—A Two-bit ADC

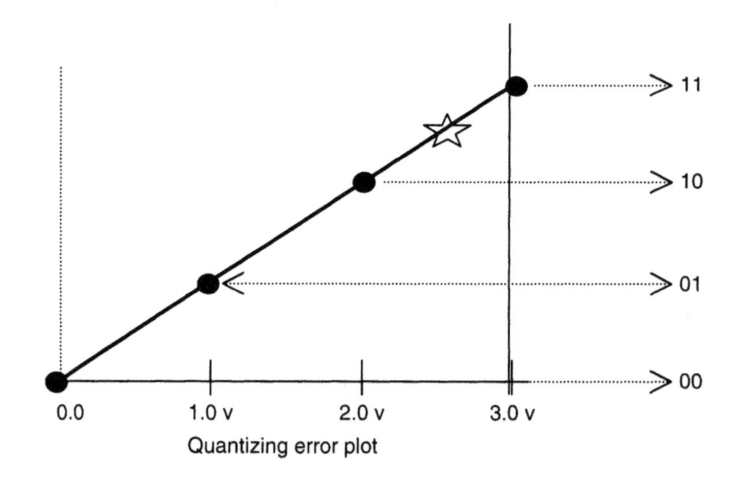

Figure 4.5

Quantizing Error Plot

4.4.2 Quantizing Error and the LSB

If we divide the full range of the analog input by the total number of available digital codes, we can derive the smallest analog step size that can be represented by a unique digital code. Because this is equivalent to the Least Significant Bit of the digital code, it is called the LSB value. Quantizing error is the uncertainty "between bits."

The worst-case quantizing error occurs when the signal value is exactly between two code thresholds and is never more than one-half of the LSB value.

$$QN < lsb/2$$

Consider a 4-bit ADC, with a 1.5-volt full-scale input range and 100-mV steps. An input level of 0.4 volts will cause the ADC to generate an output code of 0100. What will be the output code if the input level is 0.41 volts? How about 0.42 volts? You see the problem, eh? That's quantizing error.

Table 4.2

Discrete Digitizer Increments

Input Level	Output Code
1.5	1111
1.4	1110
1.3	1101
1.2	1100
1.1	1011
1.0	1010
0.9	1001
0.8	1000
0.7	0111
0.6	0110
0.5	0101
0.4	0100
0.3	0011
0.2	0010
0.1	0001
0.0	0000

The moral of the story is that the tester will lie to you. It can't help it. When we try to capture a signal, the tester will process the signal based on the acquired data, but we should not expect the acquired data to exactly match the signal. The error introduced by the digitizer process usually looks like noise. Even if the signal from the DUT is perfectly pure, the tester will actually add noise, because of the quantizing error.

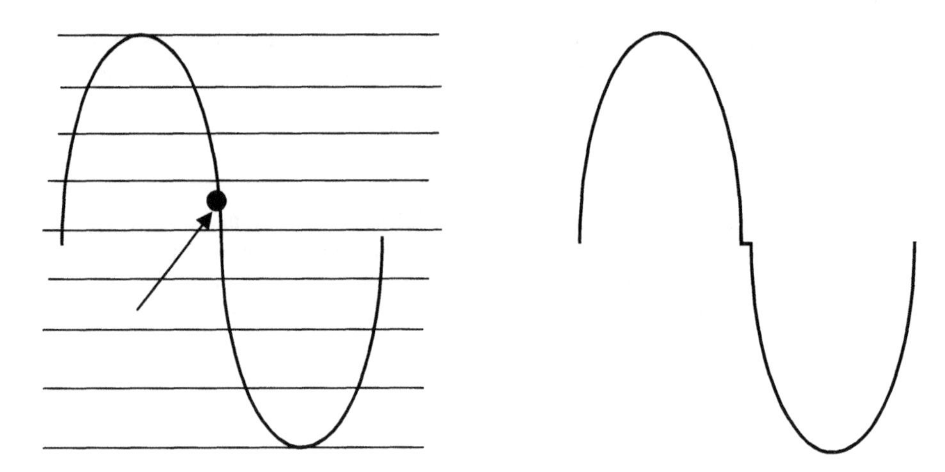

Figure 4.6

Dynamic Signal Effects of Quantizing Error

4.4.3 Digitizer Resolution

Because quantizing error is related to the LSB value, the range and resolution of the digitizer's analog-to-digital converter directly affects the amount of quantizing error. What will be the quantizing error for ADCs of different resolution with 5.0-volt signal?

Table 4.3

Increased Resolution Reduces Quantizing Error

# Bits	Total Number of Codes	LSB Value @ 5V	Quantizing Error @ 5V
7	128	39 mV	19.5 mV
12	4096	1.2 mV	0.6 mV
14	16384	305 uV	152 uV
16	65536	76 uV	38 uV

With a 5-volt range on a 16-bit capture instrument, the uncertainty is no more than 38 uV; a precision of 500 times greater than a 7-bit capture instrument.

Example

When using a 16-bit capture instrument on a 5-volt range, the instrument returns a reading of 38 uV. This may or may not be the same as the actual signal level. Because of the limitations of the digitizer, the actual level could be anywhere from 0 volts to 76 uV—there is no way to tell without improving the instrument resolution.

4.4.4 Sample Size and Sample Rate

The purpose of digitizing a signal is to construct a sample set that represents the signal amplitude over time. The continuous analog data must be sampled at discrete intervals that must be carefully chosen to ensure an accurate representation of the original analog signal. The sample frequency is the rate at which the digitizer samples the input signal for conversion to a set of discrete numeric values.

Ideally, the greater the number of samples (i.e., the higher the sample rate) that can be taken for any given signal duration, the greater the accuracy of the digital representation. However, acquiring many samples may take longer than acquiring fewer samples. Processing many samples usually takes longer than processing fewer samples. The constraints of the digitizer limit the sample frequency. The practical rule is to digitize the signal with as few samples as possible, but no fewer.

4.5 Nyquist and Shannon—Theoretical Limits

Nyquist's Sampling Limit—The sampling frequency must be greater than twice the bandwidth of the signal.

Shannon's Theorem—If a signal over a given period of time contains no frequency components greater than f_x, then all of the needed information can be captured with a sample rate of $2 \times f_x$.

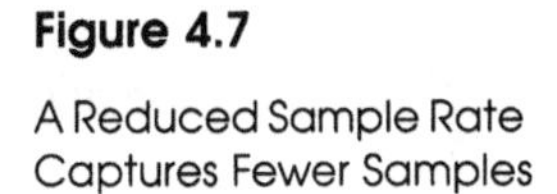

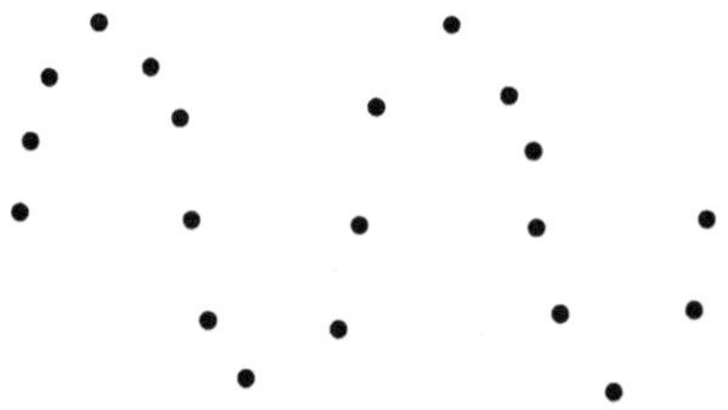

Figure 4.7

A Reduced Sample Rate Captures Fewer Samples

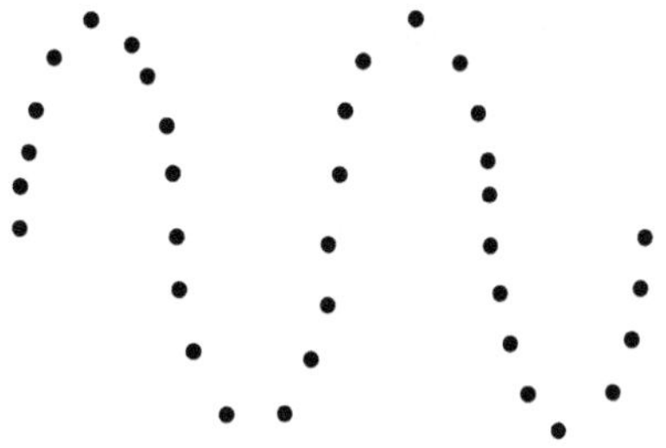

Figure 4.8

An Increased Sample Rate Captures More Samples per Cycle

4.5.1 Applications of Nyquist and Shannon

If a test program is designed to test a specific range of frequencies, the minimum sample rate may be only slightly greater than twice the bandwidth of the signal. The signal information can be *completely determined* if this condition is met. For capturing periodic signal information, just one extra sample is sufficient. At it turns out, the phase of the signal relative to the beginning and end of the capture sequence does not affect the quality of the data set. If there is an unknown phase shift in the DUT or the instrumentation, this does not prevent measuring the magnitude and frequency of the captured signal.

The underlying assumption in both Shannon's Theorem and Nyquist's Limit is that the sampling rate is consistent. That is, the sample period must be without variation across the sampling set. An *average* sample rate of $2 \times f_b$ does not meet the minimum conditions.

When testing for noise and distortion, estimating the actual signal bandwidth requires careful attention. A 1-kHz sine wave, for example, could theoretically be captured with a sample rate of 2.01 kHz. Applying Shannon's theorem, however, it becomes clear that the only data to be completely captured will be less than 1.005 kHz (fs/2). This would collect no information about the harmonics or noise above 1.005 kHz.

Example

To calculate the actual minimum sample rate requires understanding the actual signal bandwidth, including distortion and noise. What would be the minimum sample rate to properly capture the following signal data?

- Test signal is 10.000 kHz sine wave
- Test for total harmonic distortion (THD) includes the 2nd and 3rd harmonics
- Test signal-to-noise (SNR) ranging from 1 kHz to 50 Khz

Even though the primary signal is 10.0 kHz, the actual bandwidth of interest extends to 50.0 kHz. In that case, the sample frequency must be greater than 100 kHz, or twice the bandwidth of interest.

4.5.2 Signal Aliasing

What happens if Shannon's Theorem is not met? What happens if the test program samples a signal at less than twice the signal frequency?

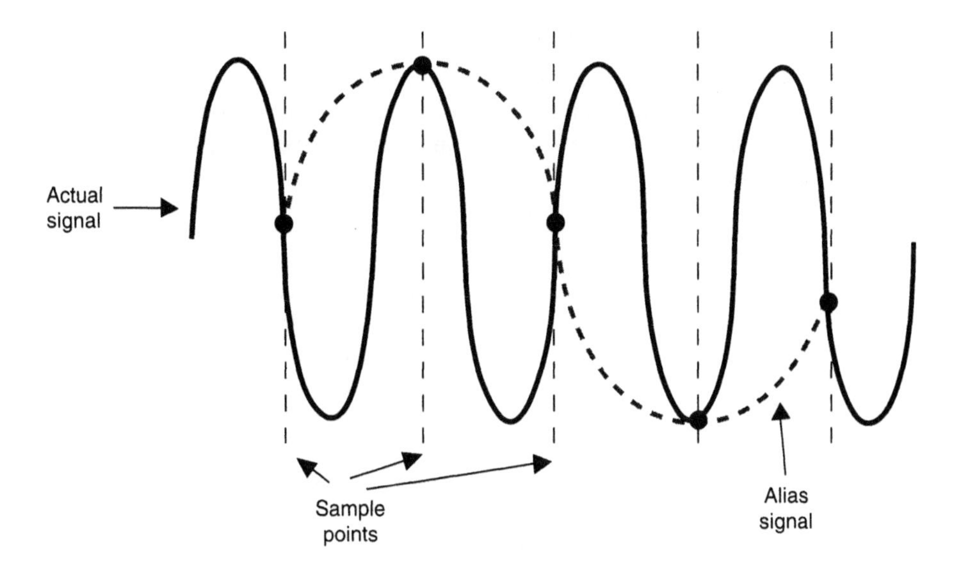

Figure 4.9

Inadequate Sampling Rate Produces Bogus Sample Set

Big trouble! If the test program captures fewer than two samples per signal cycle, a false representation of the signal is generated from the available data points. If we view only the sample point from the preceding diagram, it appears as though the signal is of much lower frequency than the actual data. The test application can only interpolate based on the available data points.

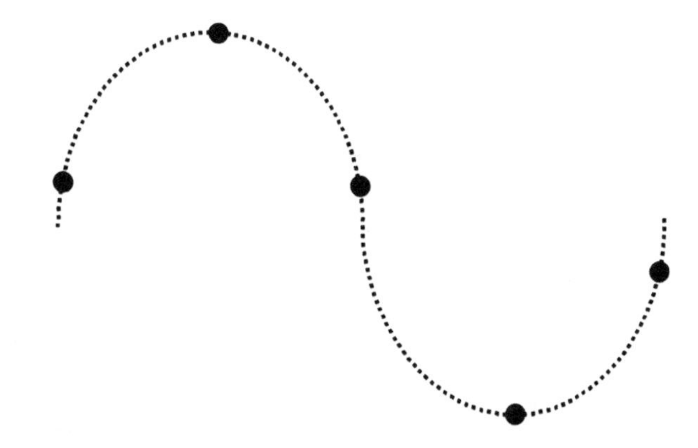

Figure 4.10

Signal Aliasing

The false signal is known as an *alias*. This error appears in both the time domain and the frequency domain. The purpose of the anti-alias filter in the capture hardware is to remove high-frequency signals that could cause an alias.

4.5.3 The Anti-Alias Filter

The roll-off characteristics of the anti-alias filter play a part in calculating the correct sample rate. A typical anti-alias filter in a signal capture unit may be configured as a selectable bank of filters. Each filter has a specified pass-band, and a roll-off characteristic.

Filter Bank Pass-Band	Filter Band Roll-off
0 Hz to 500 Hz	–24 dB per octave
0 Hz to 1000 Hz	
0 Hz to 2000 Hz	
0 Hz to 4000 Hz	
0 Hz to 8000 Hz	
0 Hz to 16000 Hz	

The roll-off specification indicates that signal frequencies above the pass-band will be rejected by 24 dB for every doubling of the signal frequency. Selecting the 0–1000-Hz filter will pass all frequencies within that range. Frequency components above the pass-band would be attenuated as follows:

Spurious Signal Frequency	Attenuation
2 kHz	–24 dB
4 kHz	–48 dB
8 kHz	–72 dB
16 kHz	–96 dB

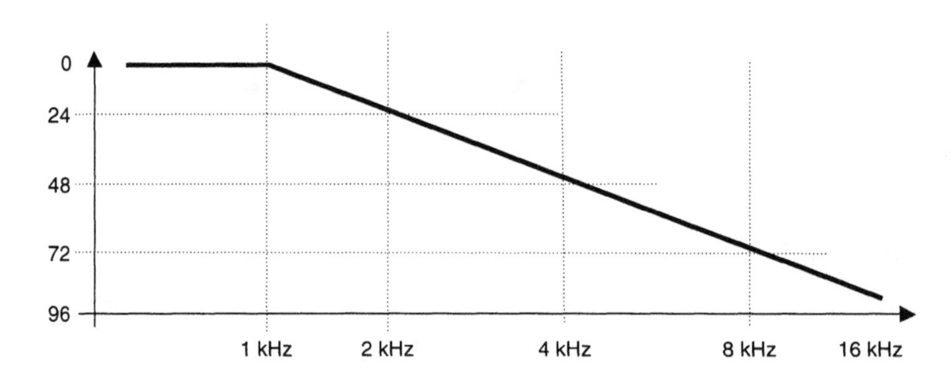

Figure 4.11

Anti-alias Filter Slope

If you wanted to make sure that all signal components above the Nyquist point were attenuated by at least 96 dB, the sampling rate would need to be 32 kHz. You must program the sample clock at a high enough frequency to place the Nyquist point at a usable point on the filter slope. It's a lot like what we went through with the reconstruction filter. First, you need to understand the filter response, then you need to calculate the sample frequency that will cause the Nyquist point to intercept the filter slope.

Filter Selection Example

Suppose you selected the 1-kHz band pass anti-alias filter. This would set up the capture hardware to accept all signal frequencies below 1 kHz, and to reject signal frequencies about 1 kHz. If the capture instrument sample rate (fs) is programmed to 16 kHz, then the Nyquist frequency would be fs/2, or 8 kHz.

Note that the selected configuration of the anti-alias filter has an attenuation of only –72 dB at 8 kHz. This would create inadequate filtering of spurious high-frequency signal information, and could generate incorrect measurement results. In order to avoid false measurements due to signal information above Nyquist, the sample frequency (fs) must be high enough to place the Nyquist point (fs/2) in the filter reject region. In order to fully utilize the anti-alias filter, it is common practice to choose a sample frequency that is at least 16 to 32 times the signal bandwidth.

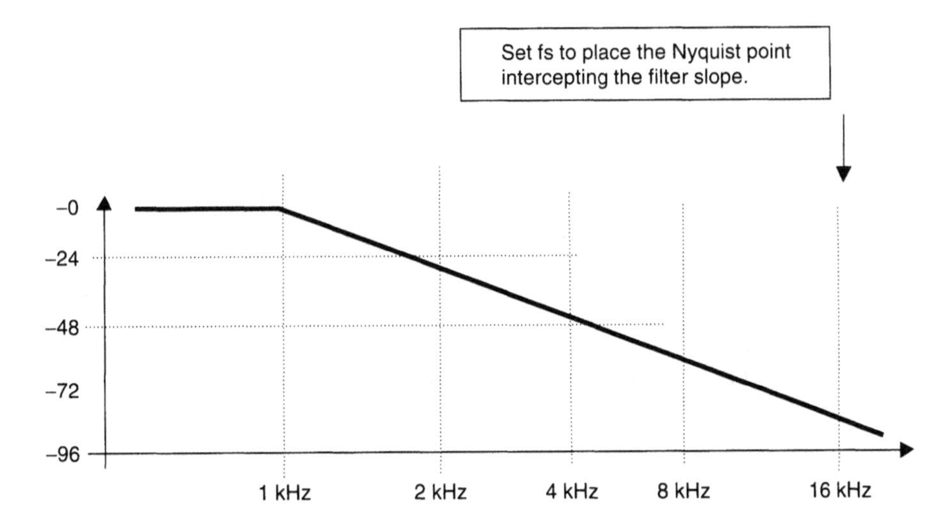

Figure 4.12

Filter Selection Example

What Was That Again?

OK, it does seem a little backward, so let's take another look. You want to make sure that you do not process any information above the Nyquist point, because of all the weird things that happen to the signal. But, we're stuck with the filters that the ATE system gives us. If we had a super-deluxe filter with a 120-dB roll-off per octave, then we wouldn't have to worry. Because the actual filter isn't really all that steep, and it does not start to seriously kick in until three octaves (every octave doubles the frequency) above the pass-band, there's only one thing left. We can't change the target frequency, and we can't change the filter roll-off—the only thing we can do is change the Nyquist point. By increasing the sample frequency, you can push the Nyquist point out to where it intersects the filter slope.

4.6 Sampling Rate and the Frequency Domain

In the last section, we saw how a multi-tone waveform can be created by combining several sine waves of different frequencies. The result was a non-sinusoidal wave shape. As it turns out, any non-sinusoidal wave shape can be analyzed as a multi-tone, that is, as a signal made up of sine wave components. Frequency domain analysis deconstructs a signal into the sine wave components, and graphs the results according to magnitude across frequency.

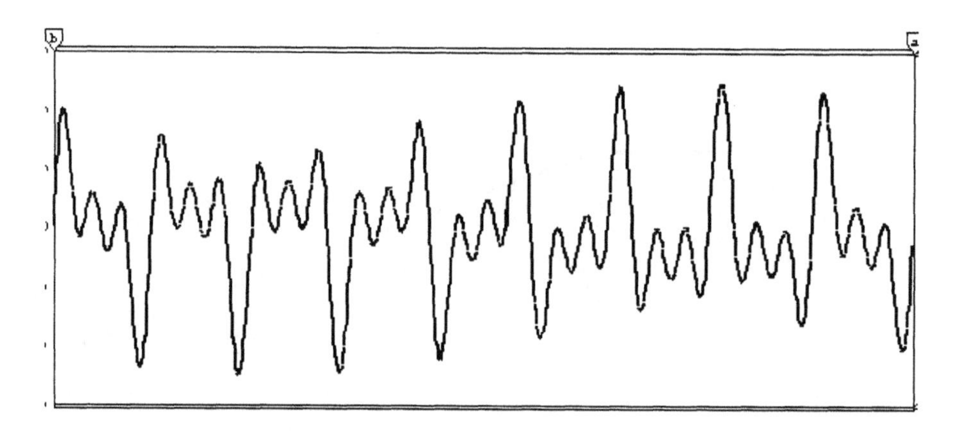

Figure 4.13

Time Domain Multi-tone

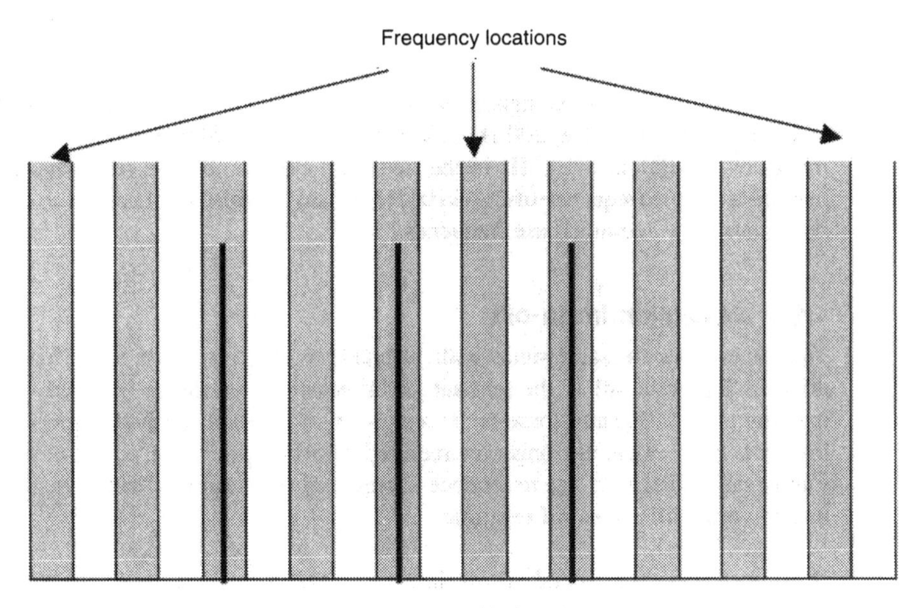

Figure 4.14

Frequency Domain Multi-tone

4.6.1 Frequency Resolution

The value of the base frequency (fbase) determines the step size in the frequency domain. In the time domain, the value of fi/fbase indicates the number of cycles in the sample set. In the frequency domain, the value of fi/fbase determines the signal location. Processing the digitized signal in the frequency domain requires that the digitizer sample rate and sample size is sufficient to produce the desired resolution. The fbase value is the step size, or x-axis resolution, in the frequency domain.

$$\text{fbase} = \text{frequency resolution}$$

Suppose that in order to test a captured signal for noise and distortion, you decide to measure the frequency domain energy at 100 Hz, 200 Hz, 300 Hz, 400 Hz, and 500 Hz. This means you need frequency domain data in increments of 100 Hz. To achieve a frequency resolution of 100 Hz, the sample size and sample rate must be selected to generate the corresponding fbase value.

$$\text{fbase} = \frac{\text{fs}}{\text{samples}}$$

There's no DSP magic that can bail you out here. The *only* way to establish the proper resolution in the frequency domain is to choose the correct sample rate and sample size in the time domain.

Example

In order to test a captured signal for noise, you decide to measure the frequency domain energy at 100 Hz, 200 Hz, 300 Hz, 400 Hz, and 500 Hz. You decide on a frequency resolution of 100 Hz in the frequency domain, but the test specification requires a sample frequency of 25.6 kHz. How many samples must be in your sample set to derive the required base frequency?

4.6.2 Resolution Trade-offs

You can evaluate the same signal with different levels of resolution in the frequency domain. The trade-off is the amount of information versus the required capture duration period. Because fbase is the reciprocal of the capture duration period, the lower the fbase value, the longer the required duration period. You could set up your sample rate and sample size to produce a frequency resolution of 1 Hz, but it's going to cost you a full second of test time.

How much does one second of test time cost? Well, if you've working with a new tester platform, it could cost around 4 million dollars. By the time you factor in the amortization, interest, and utilization factors, the equipment cost works out to more than a million dollars per year. There's about 31 million seconds in a year, so we're looking at a tester cost of 3.3 cents per second.

Let's look at one more example before we move on.

Parameters for 250-Hz Frequency Resolution	
1250-Hz signal frequency 32-kHz sample frequency	250-Hz frequency resolution 128 samples

Result: The frequency domain data set will be arranged in steps of 250 Hz. That is, the frequency domain will show data at 0 Hz, 250 Hz, 500 Hz, 750 Hz, etc.

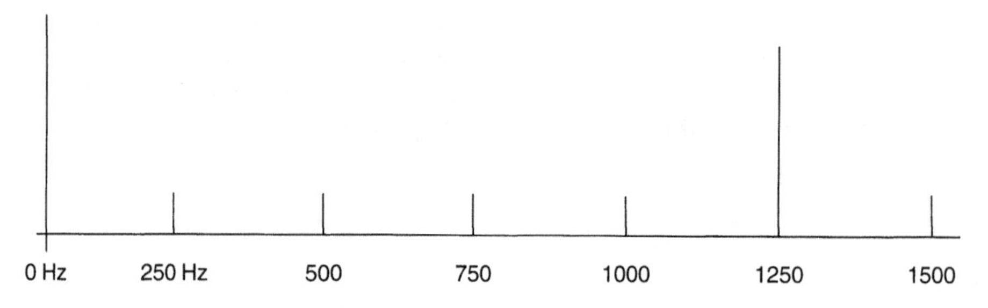

Figure 4.15

250-Hz Frequency Resolution

Parameters for 50-Hz Frequency Resolution	
1000-Hz signal frequency	50-Hz frequency resolution
51.2-kHz sample frequency	1024 samples

Result: The frequency domain data set will be arranged in steps of 50 Hz. That is, the frequency domain will show data at 0 Hz, 50 Hz, 100 Hz, 150 Hz, etc.

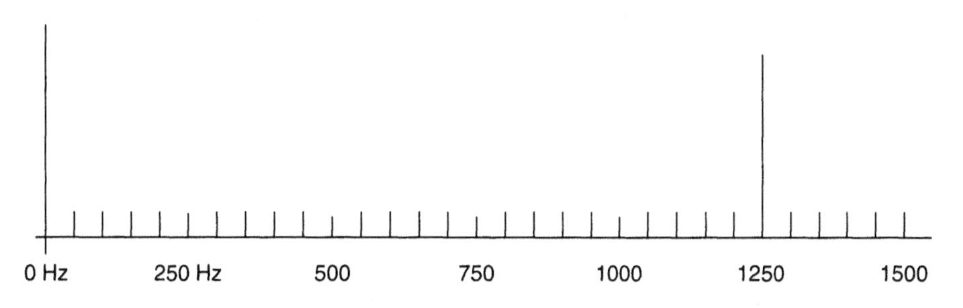

Figure 4.16

50-Hz Frequency Resolution

4.7 Capturing Periodic Sample Sets

Digitizing an analog signal has many of the same requirements as synthesizing a signal. Generating a useful data set for the analog signal source requires calculating the correct sample rate and sample size to produce an integer number of cycles. Acquiring a useful data set for the analog signal capture also requires calculating the correct sample rate and sample size to obtain a periodic sample set.

The *sample window* refers to the time period during which the digitizer is sampling the analog data, and is calculated by

$$\text{sample window duration} = \text{number of samples} \times \text{sample period}$$

The window must be of the proper duration to correctly capture a periodic sample set of the analog signal. If the sample set duration is too long or too short, the captured signal data set will not properly represent the analog data.

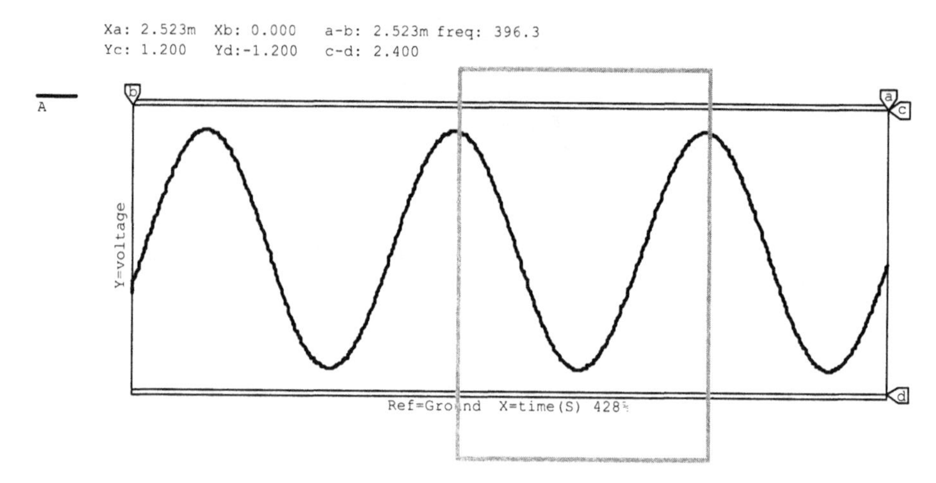

Figure 4.17

Capturing a Periodic Sample Set

Testing for frequency domain characteristics often requires the use of a computer algorithm called the Fast Fourier Transform (FFT). Once of the constraints of the FFT is that the input data will be processed *as if it were* periodic. The design of the FFT algorithm assumes that the data set can be duplicated without introducing signal error.

4.7.1 Application Example Overview

Let's work through an example to illustrate the process for calculating capture instrument parameters.

OBJECTIVE: Determine the duration of the sample window that will capture an integer number of cycles for each frequency component of the analog signal.

PARAMETERS: The duration of the sample window is determined by the sample size and the sample rate.

PROCESS: Calculate the frequency resolution that divides evenly into all of the frequency components of the analog signal. Once the fbase has been determined, you can derive the proper sample size and sample rate.

Our objective is to calculate the sample size and sample rate for digitizing a multi-tone signal composed of 1680 Hz, 3750 Hz, and 5460 Hz.

4.7.2 The Largest Common Denominator

Step One: Remember that fbase is a periodic frequency, which represents the ratio of the sample frequency over the sample size. What is the periodic frequency, or fbase, that will divide evenly into all three components? (Find the largest common denominator.)

$$\text{Factoring } 1680 = 2 \times 2 \times 2 \times 2 \times 3 \times 5 \times 7$$
$$\text{Factoring } 3570 = 2 \times 3 \times 5 \times 7 \times 17$$
$$\text{Factoring } 5460 = 2 \times 2 \times 3 \times 5 \times 7 \times 13$$
$$\text{Common factor} : 2 \times 3 \times 5 \times 7$$

The largest common denominator is therefore 210 Hz, which we will select as the fbase. Recall that fbase is the reciprocal of the window duration.

$$\text{window duration} = \frac{1}{\text{fbase}} = \frac{1}{210 \text{ Hz}} = 4.7619 \text{ mS}$$

4.7.3 Verify Integer Number of Cycles

Step Two: Given an fbase frequency of 100 Hz, verify that the captured data set will contain an integer number of cycles for each component of the multi-tone.

$$\text{cycles} = \frac{\text{fi}}{\text{fbase}}$$

1. Tone One = 1680 Hz

$$\frac{\text{fi}}{\text{fbase}} = \frac{1680}{210 \text{ Hz}} = 8 \text{ cycles}$$

2. Tone Two = 3750 Hz

$$\frac{\text{fi}}{\text{fbase}} = \frac{3750 \text{ Hz}}{210 \text{ Hz}} = 17 \text{ cycles}$$

3. Tone Three = 5460 Hz

$$\frac{\text{fi}}{\text{fbase}} = \frac{5460 \text{ Hz}}{210 \text{ Hz}} = 26 \text{ cycles}$$

That is, the 4.7619-mS capture duration window will contain an integer number of cycles for each sine wave frequency of the multi-tone waveform.

4.7.4 Approximate fs and Sample Size

Step Three: Once you capture a periodic sample set of the multi-tone signal, you will perform a Fast Fourier Transform (FFT) to analyze the data in the frequency domain. The FFT transform requires that the sample set be a power of two, so you need to choose the number of samples and the sample rate to produce

1. an fbase of 210 Hz
   AND
2. a power of 2 sample size (i.e., 32, or 64, or 128...)

We also know that the sample frequency should be at least 16 times the signal bandwidth, in order to make the best use of the anti-alias filter. As an approximation, we can therefore estimate the sample frequency as

$$fs \approx f_b \times 16 = 5460\text{ Hz} \times 16 = 87360\text{ Hz}$$

Step Four: Applying DSP's law, we can approximate the sample size from the preliminary sample frequency value, and the selected fbase value.

$$fbase = \frac{fs}{samples} \rightarrow samples = \frac{fs}{fbase} = \frac{87360\text{ Hz}}{210\text{ Hz}} = 416\ \text{ samples}$$

The only problem is that the FFT requires a power of two sample size. That is, the FFT can process a sample size of 2, or 4, or 8, or 16; but it cannot process a sample set that is 5, or 14, or 18. Because the preliminary sample size of 416 samples is not a power of two, we must go through another iteration to determine the exact values for the sample frequency and sample size parameters.

4.7.5 Optimizing Capture Parameters

Step Five: We know the sample size must be greater than or equal to 416 samples to achieve the target capture parameters. To meet the FFT data set requirement, we must choose a sample size that is a power of two, and is also greater than 416. Recalculating with a sample size of 512 samples gives us the following:

$$fbase = \frac{fs}{samples} \rightarrow fs = samples \times fbase = 512 \times 210\text{ Hz} = 107520\text{ Hz}$$

Step Six: As a proof, let's apply the Golden Ratio to this application. Recall that the ratio of the sample frequency (fs) to the signal frequency (fi), is always the same as the ratio of the number of samples to the number of cycles.

fs:fi = samples:cycles

$$\frac{fs}{fi} = \frac{107520\ Hz}{5460\ Hz} = 19.6923 \rightarrow \frac{samples}{cycles} = \frac{512}{26} = 19.6923$$

Now, is that cool, or what?

4.8 Signal Averaging

Signal averaging is a method for cancelling random noise in the captured signal data set. Over time, the value of Gaussian distribution noise averages to zero; so by taking an average of several signal cycles, the random noise error can be removed.

Consider five signal cycles, each with a random noise component. By summing the signal data and then generating an average, the random noise components are reduced by a factor equal to the square root of the number of averaged cycles. Because the signal information is periodic, the amplitude of the averaged data is unity for the signal. Because the noise components are non-periodic, the averaged data attentuates the noise information.

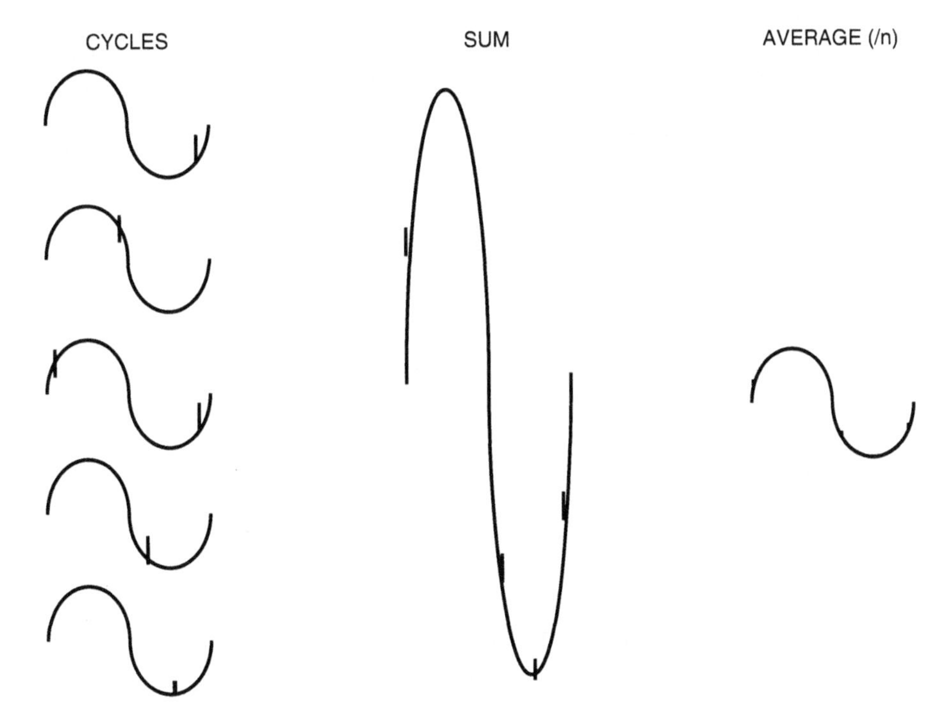

Figure 4.18

Signal Averaging Reduces Random Noise

4.9 Capturing Unique Data Points

In general, most mixed signal test applications will return more complete results if the captured data is not redundant. For example, if the test program captures a 1.0-kHz signal using a sample rate of 16.0 kHz, each cycle of the signal will be represented by 16 consecutive samples. If the test program collects 5 cycles, it will have gathered the same 16 data points, relative to the signal cycle, 5 consecutive times.

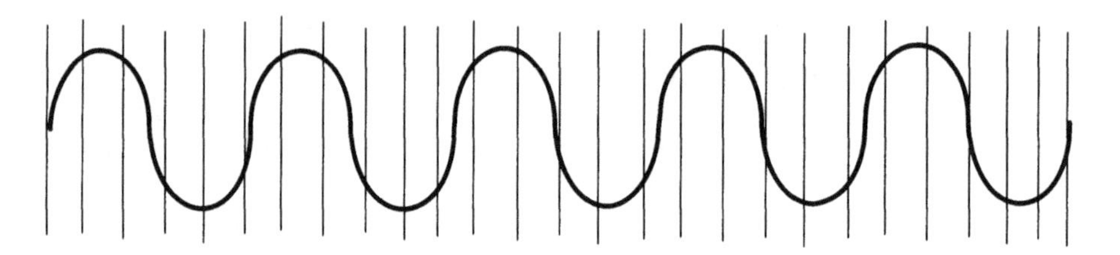

Figure 4.19

Synchronous Samples Produce Redundant Data

To collect new data for each cycle, the sample frequency and the signal frequency must be mutually prime. An integer ratio of fs/fi results in a redundant data set when capturing more than one cycle. A non-integer ratio of fs/fi allows each sample in the captured data set to represent a unique point of the captured signal.

This is one method for calculating the correct fs and sample size to ensure unique data points in the sample set.

1. Choose a prime number cycle count.
2. Recalculate

$$fs = fs \times \left(\frac{1}{cycles} + 1 \right)$$

3. Recalculate the sample size as

$$samples = \left(\frac{fs}{fbase} \right)$$

4.9.1 Applications Example

Step One: The first iteration has the following parameters:

fs = 16 kHz

fi = 1 kHz

fbase = 200 Hz

samples = 80

cycles = 5

Relative to the signal, this will collect the same 16 data points for each cycle. In order to capture unique data points for each cycle, we need a second iteration of calculations.

Step Two: Multiply fs by the reciprocal of the number of cycles +1.

$$fs = fs \times \left(\frac{1}{\text{cycles}} + 1 \right) = 16000 \text{ Hz} \times \left(\frac{1}{5} + 1 \right) = 19200 \text{ Hz}$$

Calculate the new sample size:

$$\text{samples} = \frac{fs}{\text{fbase}} = \frac{19200 \text{ Hz}}{200 \text{ Hz}} = 96 \text{ samples}$$

With these new parameters, the ratio of fs to fi is a non-integer, as is the number of samples per cycle.

$$\frac{fs}{fi} = \frac{19200 \text{ Hz}}{1000 \text{ Hz}} = 19.2 \text{ samples per cycle}$$

The number of signal cycles has not changed, only the number of samples per cycle. As a result, the capture duration, and therefore the signal acquisition time, is the same even though more information is being gathered.

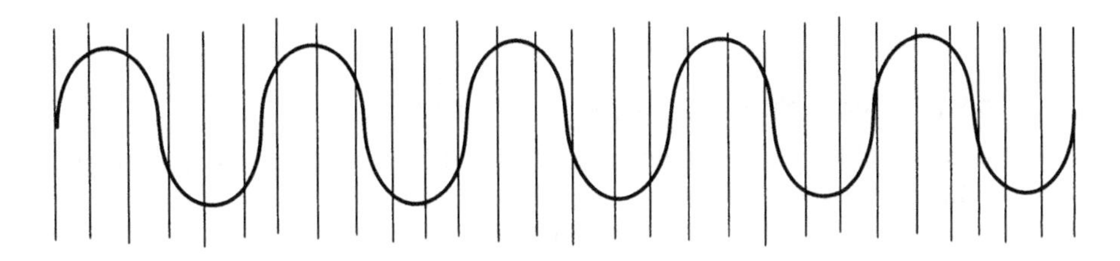

Figure 4.20

Mutually Prime fs and fi Produce Unique Samples

4.9.2 Over-Sampling Technique

Another application of capturing unique sample points concerns a method known as *over-sampling*. In applications where the sample per cycle ratio is small, the digitizer can be programmed to capture unique points on the signal for each cycle. The end result is an effective sample rate, that is higher than the actual fs.

This technique is similar to that used in sampling scopes, where the signal is reconstructed on the basis of data collected under a large number of cycles. In order to properly reconstruct a signal from data collected under several cycles, the samples must take place at different points for each cycle, relative to the signal.

Consider the following scenario. You are developing a test program that includes a requirement to capture a 200.0-kHz signal at 10.0-MHz sample rate. The constraints of the ATE system capture instrumentation are such that the maximum sample clock (fs) is 1.0 MHz. If the test objective can be met by capturing data at 100-ns increments, or an effective 10.0-MHz capture rate, then over-sampling can provide a solution.

1. Calculate the period of the over-sampled fs.
   Example: If you need an effective over-sample clock rate of 10 MHz, the period is 100 ns.
2. Add the over-sampled period to the actual sample period.
   The actual sample clock is 1 MHz, which is equal to a period of 1000 ns. Adding the over-sample period gives 1100 ns, or a frequency of 909.0909... kHz. We can round this off to 909 kHz. This is your new actual fs.
3. Use the Golden Ratio to determine the new sample size and number of cycles.

   fs:fi = samples:cycles = 909 kHz:200 kHz = 909:200 = 909 samples and 200 cycles

4. Capture unique data points across multiple cycles, and generate a composite data set.
   By capturing a large number of samples over multiple cycles, the captured data can be reconstructed as a composite 100-nS sample set.
   Note: A technique known as under-sampling is described in Chapter 5.

4.9.3 Over-Sampling Application Example

With a 200-kHz signal frequency and a 1.0-MHz sample clock, the digitizer will collect 5 samples for each cycle. By changing the sample clock frequency to 909 kHz, the digitizer will collect 4.545 samples per cycle. A total of 909 samples from 200 signal cycles is acquired and processed into a composite signal by the DSP.

For the first cycle, the digitizer collects four samples. On the second cycle, the digitizer also collects four samples, but the the sample points are 100 ns later into the signal cycle. The four samples collected from the second cycle are superimposed on the data from the first cycle. On the third cycle, the digitizer will grab another four samples, which are now 200 ns later into the signal cycle. The over-sampling technique collects lots of samples across lots of cycles. Because the sample rate (fs) is "out-of-sync" with the signal frequency, each sample occurs at a unique point on the signal cycle

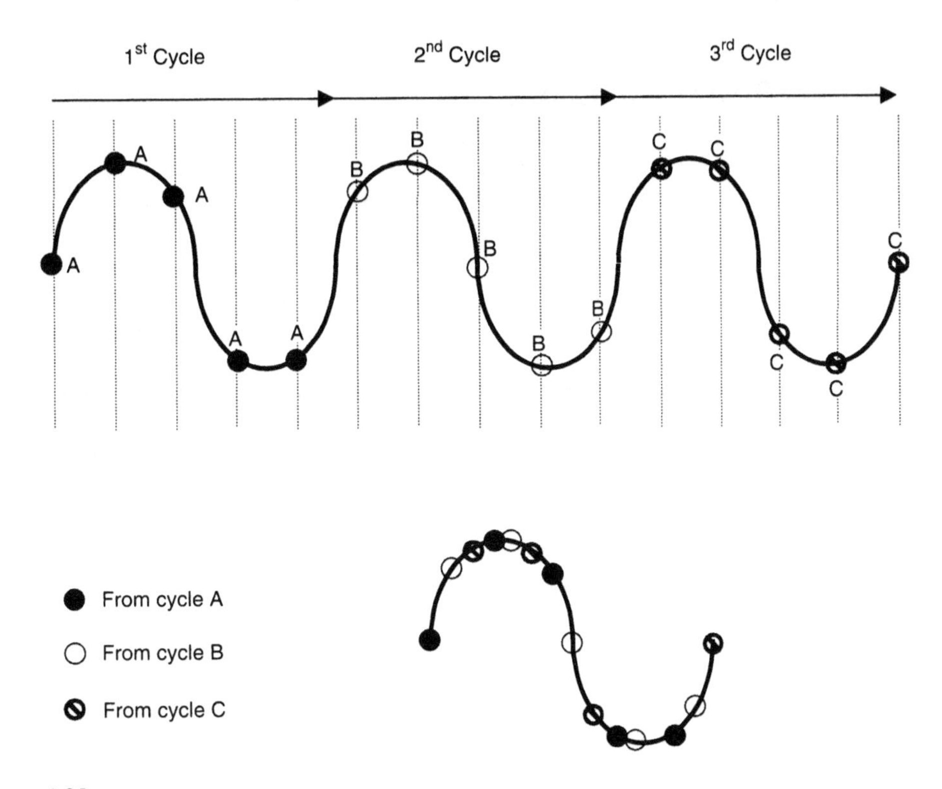

Figure 4.21

Over-Sampling Compiles Data Across Many Cycles

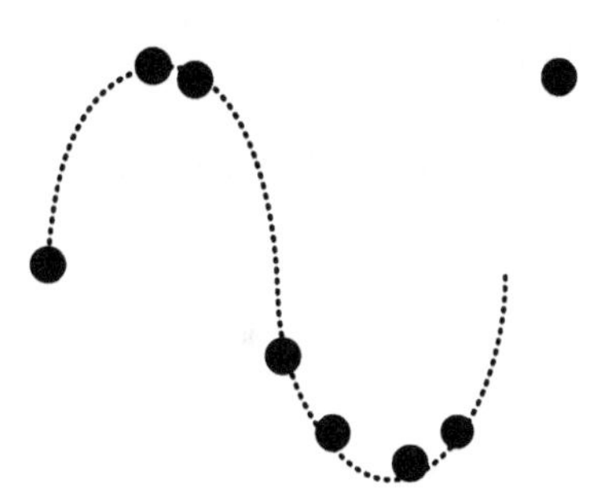

Figure 4.22

Composite from 2 Cycles

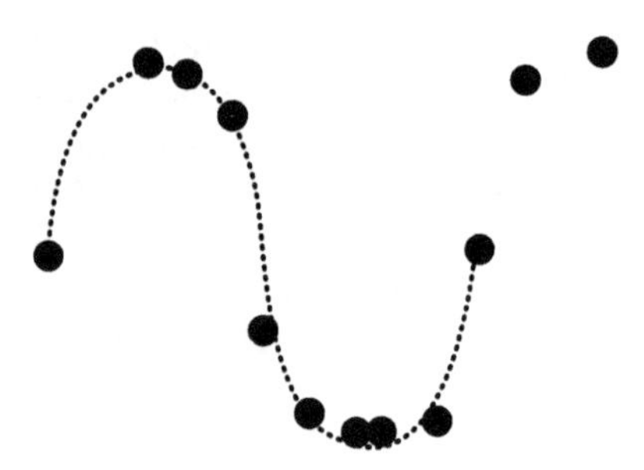

Figure 4.23

Composite from 3 Cycles

Chapter Review Questions

1. What is meant by quantizing error? What is the relationship between the quantizing error and the digitizer LSB?

2. You have an application that requires testing a 1000-Hz signal for harmonic distortion and noise. You decide that to test for noise, you will measure the signal energy at every 100 Hz, and that you will capture 512 samples.

   What is the fbase (base frequency)? ___________

   What is the digitizer frequency (fs)? __________

3. What is the importance of fbase in the frequency domain?

4. The device under test generates a sine wave signal of 1500 Hz. You choose a sample size of 128, and a sample frequency of 64000 Hz.
   What is the fbase? _________

   How many cycles of the signal of interest (1500 Hz) will be in the sample set?

   How many samples will there be for each cycle of the signal? _________

CHAPTER 5

FREQUENCY DOMAIN TESTING AND THE FFT

A problem worthy of attack shows its worth by fighting back.
—Paul Erdos, *The Man Who Loved Numbers*

5.1 Introduction

I'm about as dumb as a person can be, and even I can use the FFT. Breathe deep and relax! It's true that the innards of the Fourier transform involve some pretty heavy math, but that part has already been done. We only want to *use* the FFT; we don't need to design it. For the most part, we are going to treat the FFT like a "black box." And, I'm going to take some liberties with generalizations. If you really want to know the details about how the Fourier transform works, there are some really good books that will walk you through the process. For our purposes, we can rely heavily on over-simplification.

The Fourier transform is a math process that converts data from the time domain (such as on an oscilloscope) to the frequency domain (such as a spectrum analyzer). The fast Fourier transform (FFT) is an efficient computer algorithm for executing the Fourier transform process. The FFT is used to extract frequency domain information for tests including SNR (signal-to-noise ratio), harmonic distortion, and frequency response. The FFT is also used for test techniques such as mathematical over-sampling and digital filters. In order to use the FFT, you should understand the input requirements, the process constraints, and the output format.

Signal data captured by the digitizer is in the time domain, and forms the input data set to the fast Fourier transform. The quality of the captured data directly affects the result of the information produced by the FFT. Time domain data represent variations in amplitude over time, while frequency domain data represent variations in amplitude over frequency.

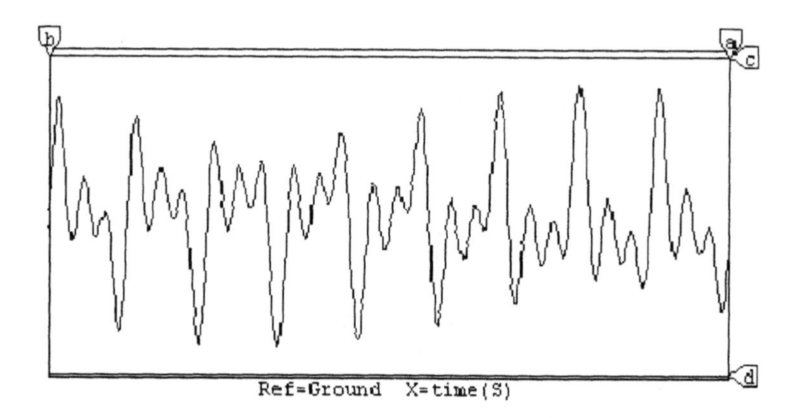

Figure 5.1

Three Tones in the Time Domain

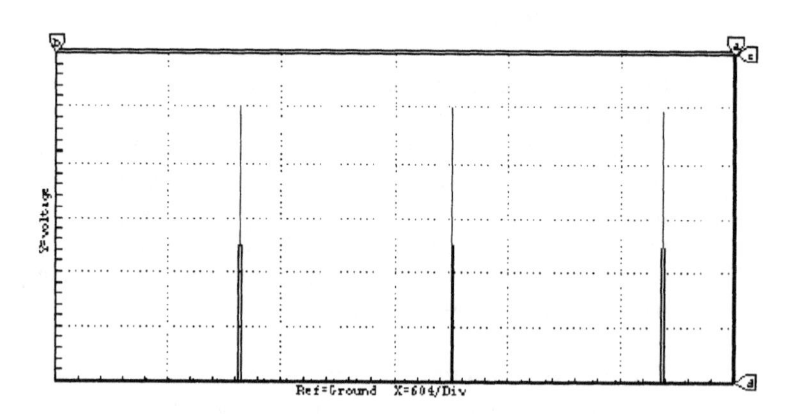

Figure 5.2

Three Tones in the Frequency Domain

5.2 The Fourier Series

One way of summarizing Fourier's theory is that any periodic non-sinusoidal waveform can be shown to be composed of a combination of sine waves, which may vary in amplitude, frequency, and phase. Any non-sinusoidal waveform is actually a bunch of sine waves trying to get out.

So, who was this guy Fourier? Jean Baptiste Joseph Fourier was a 18th-century diplomat, historian, and mathematician. Fourier's interest in heat conduction led him to begin work, in 1807, on a theory known as *The Analytic Theory of Heat*, which was published in 1822. The method of analysis has developed into a cornerstone for applications that require transforming time domain data to the frequency domain, and vice versa.

- If a waveform can be completely described with respect to TIME, then that waveform can be completely described with respect to FREQUENCY.
- If a waveform can be completely described with respect to FREQUENCY, then that waveform can be completely described with respect to TIME.

The two primary tools developed by Fourier are the Fourier series and the Fourier integral. The Fourier series is used to calculate complex frequency spectrum data based on periodic time domain data, or calculate periodic time domain data based on complex frequency domain data. The Fourier integral is used to obtain complex frequency spectrum data based on non-recurring (non-periodic) time domain data.

As it turned out, Fourier's fellow mathematicians didn't understand what he was trying to do, and publicly denounced his work. One colleague went so far as to say, "Divergent series are a work of the devil; and it is shameful to use them for any purpose whatsoever!" (quoted by Robert W. Ramirez, *The FFT, Fundamentals and Concepts*) Practically speaking, the Fourier series operates as a convergent series, but that wasn't clear at the time. Fourier must have been a pretty interesting guy, with way too much time on his hands. He was a sidekick to Napoleon in the diplomatic corps, and wound up being stationed in Egypt. While in Egypt, Fourier became the first European to make a systematic study of ancient Egyptian history. I understand that he had a specific interest in the sophisticated plumbing of that ancient era, which made him a "Pharaoh Faucet Major."[1] (Sorry, I can't help myself from writing a bad pun. It's an addiction!)

Mixed signal testing focuses primarily on the Fourier series rather than the Fourier integral, and the analysis of periodic data. When using the Fourier series in mixed signal applications, we can use data sets that comply with the Dirichlet requirements: "A finite number of minima and maxima; a finite number of discontinuities; and integrable in any period." Which is to say, a periodic sample set.

[1] Apologies to Mary Farrah Leni Fawcett.

5.2.1 The Square Wave Example

By summing together a combination of sine waves, you can approximate a square wave signal. The "recipe" for a square wave consists of summing a fundamental frequency with odd-order harmonics, in odd-order amplitude decrements. For example, a 1000-Hz square wave can be constructed from the following:

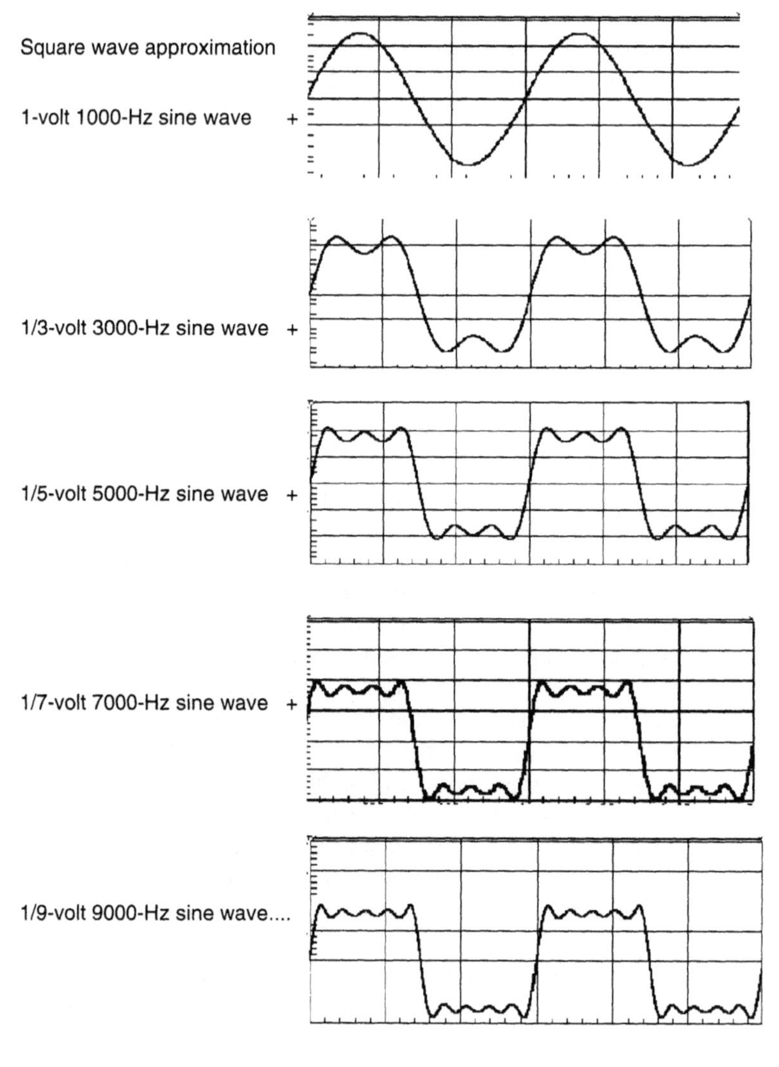

Figure 5.3

Approximating a Square Wave by Combining Sine Waves

Granted, that's not the best looking square wave in the world, but you can see where we're going. The Fourier series reverses this process, and deconstructs a signal into sine wave components. The deconstructed signal information is then organized according to the amplitude for each sine wave component. The frequency domain plot for a square wave signal, therefore, would exhibit sine wave components according to the square wave "recipe."

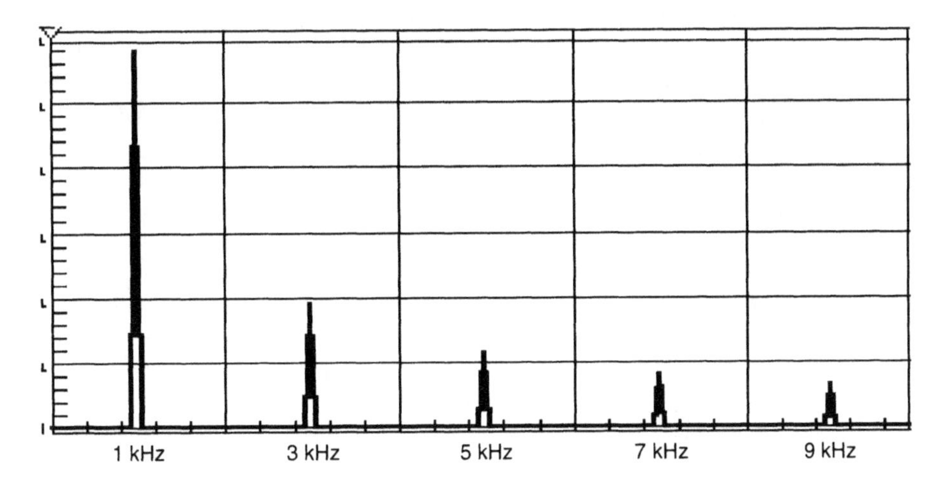

Figure 5.4

The Square Wave in the Frequency Domain

The Fourier transform performs mathematically what a spectrum analyzer instrument performs electronically. In mixed signal test applications, we can use the Fourier series as a extremely useful function that accepts time domain data and produces frequency domain data, or vice versa.

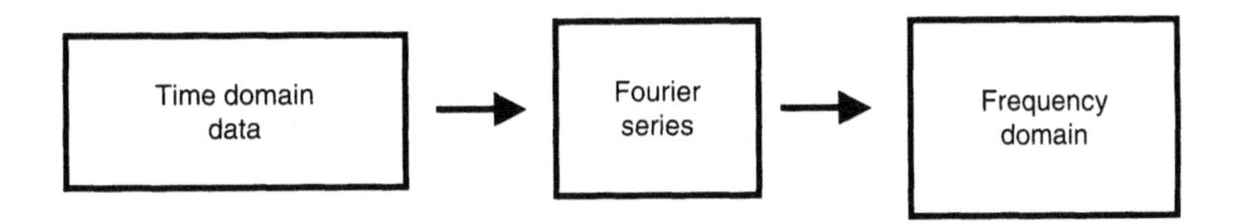

Figure 5.5

The Operation of the Fourier Series

5.2.2 The Fourier Series Equation

The actual process of the Fourier series uses a trigonometry sequence that sums the sine and cosine coefficients of the data set in sub-multiples of the period. The Fourier series as a sequence can be written as in Fig. 5.6.

Figure 5.6

The Fourier Series Equation

$$x(t) = a_0 + \sum_{n=1}^{\infty}\left[a_n \cos n\left(\frac{2\Pi}{t}\right) + b_n \sin n\left(\frac{2\Pi}{t}\right)\right]$$

Let's say that the period, $x(t)$ is equal to 1 ms. The frequency is equal to 1/t or 1000 Hz. For a period of 1 ms, therefore, the Fourier series will multiply the data set times the sine and cosine for multiples of the periodic frequency (1000 Hz, 2000 Hz, 3000 Hz, and so on). In this illustration, a_n and b_n are the coefficients from the sine and cosine products from multiples of 1000 Hz. The factor a_0 represents the DC component of the signal data set.

The Fourier series, or discrete Fourier transform (DFT) acts like a tunable bandpass filter in increments of the base frequency. In our example, the DFT would be "tuned" to DC, then 1 kHz, then 2 kHz, 3 kHz, and so on. For the first step of the process, the DFT is tuned to DC. If there are any components in the data set that correspond to DC, then the function results in a non-zero product. If there is no DC component, the function product is zero. The next steps in the process "tune" the DFT in increments of the periodic frequency. The operation multiplies the data set with the sine and cosine of the periodic frequency multiple. The output is the product of that multiplication.

5.2.3 The Discrete Fourier Transform (DFT)

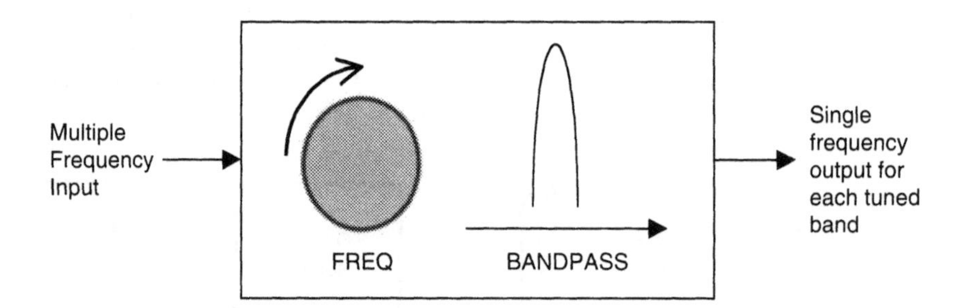

Figure 5.7

The Discrete Fourier Transform (DFT) Acts as a Tunable Filter

By multiplying the time domain data set with the sine and cosine coefficients, the discrete Fourier transform functions as a bandpass filter. For each frequency increment, the DFT effectively "passes" only that frequency. The incremental frequency steps of the DFT can be illustrated as a process that generates a sequence of amplitude values for a set of discrete frequency components. This illustration shows a very sharp and rectangular bandpass function, which is an oversimplication. We'll cover more details about actual shape of the filter a little later.

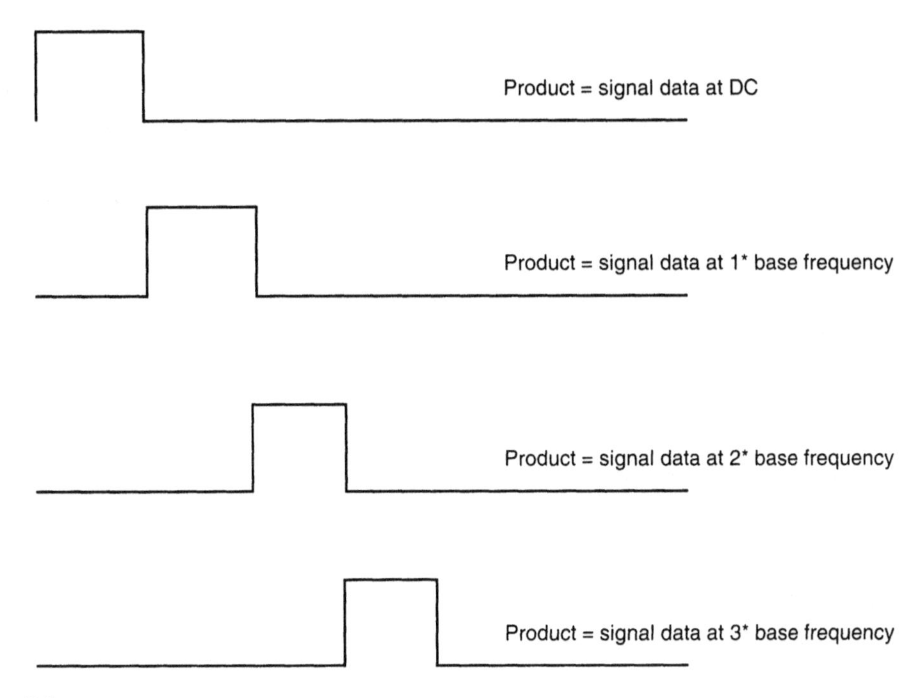

Figure 5.8
The Operation of the DFT

5.3 Representing Signal Data

5.3.1 Time Domain Data

Information that represents variations in amplitude over time is called time domain data. An oscilloscope displays time domain data. Information that represents variations in amplitude over frequency is called frequency domain data. A spectrum analyzer displays frequency domain data. Time domain data is usually processed as a sequential set of discrete sample values. The amplitude information is contained in the sample values, and the time information is inferred by the sample rate. A complete description of time domain data is actually made up of two components: time and amplitude (shown as the x- and y-axis in Fig. 5.9).

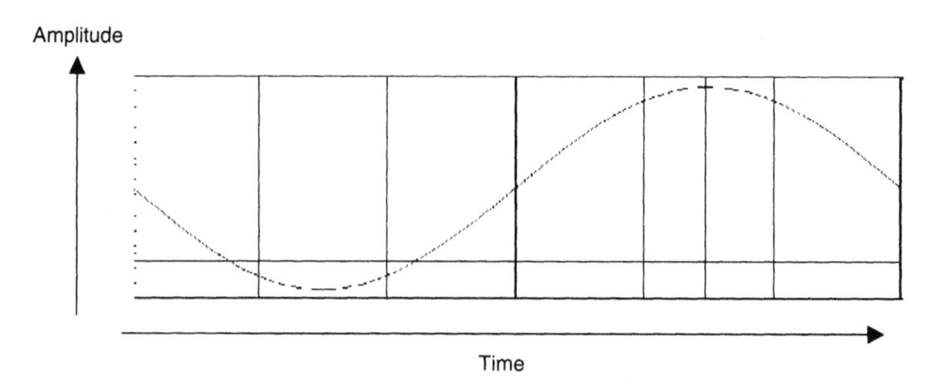

Figure 5.9

Time Domain Signal in the *x*- and *y*-axis

y	0.0	–0.3	–0.5	–0.7	–1.0	–0.7	–0.5	–0.3	0.0	0.3	0.5	0.7	1.0	0.7	0.5	0.3
x	0	1	2	3	4	5	6	7	8	9	10	11	12	13	14	15

Figure 5.10

Time Domain Signal Data Representation

The mathematical representation of a time domain signal is functionally a two-dimensional array. However, because the *x*-axis is sequential, time domain data is typically stored and processed as a single dimension array. The *x*-axis informaiton is implied from the array element location.

5.3.2 Frequency Domain Data

Like the two-dimensional array for time domain data, frequency domain data is described with a two-dimensional format. The mathematics term for this format is a **complex** number. The two components are referred to as *real* and *imaginary.* If we graph the frequency domain data on vertical and horizontal axes, the two-dimensional data (real and imaginary) is said to be in a *rectangular coordinate* system.

For the purposes of our discussion, we can view frequency domain data being made of two components, like the two-dimensional array for time domain data. Frequency domain data is described with a special two-dimensional format. The mathematics term for this format is a complex number. The two components are referred to as real and imaginary. If we graph the frequency domain data on vertical and horizontal axes, the two-dimensional data (real and imaginary) is again said to be in a rectangular coordinate system.

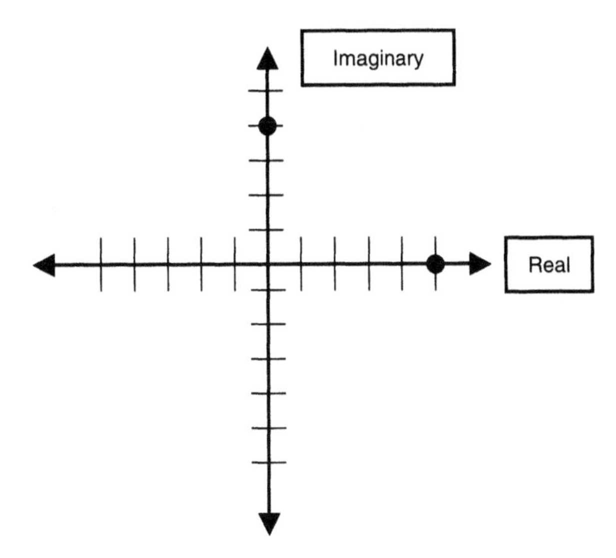

Figure 5.11

Frequency Domain Signal as Real and Imaginary Components

Don't panic! Each element in the complex frequency domain data set has two components, the real and the imaginary, just as each element in the time domain data set has both a time and amplitude component. For the purpose of our discussions, we are going to treat complex data as a two-dimensional array, nothing more.

Real	0.0	2.3	0.5	0.7	0.3	1.2	0.5	3.4	2.7	1.2	0.5	0.7	1.0	0.7	0.5	0.3
Imaginary	0.0	1.2	0.2	3.1	0.7	5.3	0.2	3.1	0.7	5.3	0.3	1.2	0.5	3.4	0.3	1.2

Figure 5.12

Frequency Domain Signal Data Representation

5.3.3 Output Format—Rectangular and Polar

If we "connect the dots" in the rectangular graph, we have a right-angle triangle. By applying some simple geometry, we can extract a single value that represents the *magnitude* of the data. We can also extract phase information. This format is called a polar coordinate system.

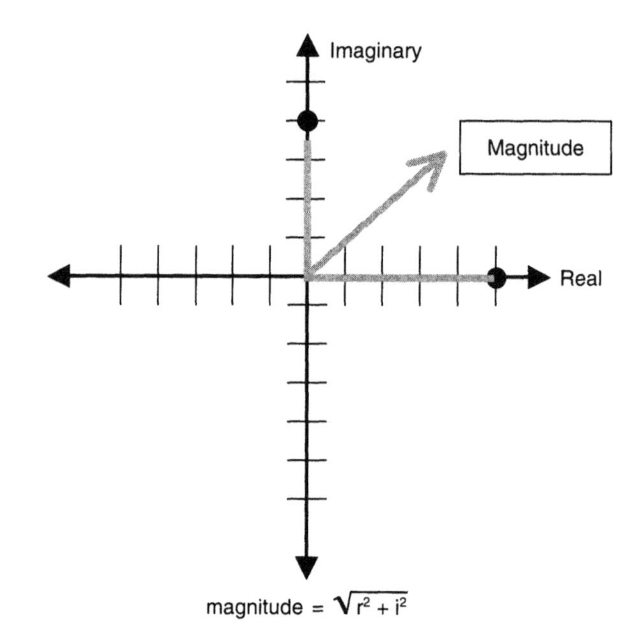

Figure 5.13

Rectangular and Polar Formats

The y-axis of the conventional frequency domain plot indicates the relative amplitude of each processed frequency component. The units of the y-axis depend on the system software, but usually are correlated to relative voltage energy (RMS) or voltage peak.

5.4 The Inverse Fourier Transform

One of the implications of the Fourier transform is that signal information can be processed in either the time domain or the frequency domain. The process can be reversed with an inverse Fourier transform, which converts data from the frequency domain into the time domain. Some test and signal processing techniques such as brick-wall filters and mathematical over-sampling process the signal in the frequency domain, and then convert the data back into the time domain using an inverse Fourier transform. The Fast Fourier Transform is abbreviated as the FFT, and the inverse is known as the IFFT.

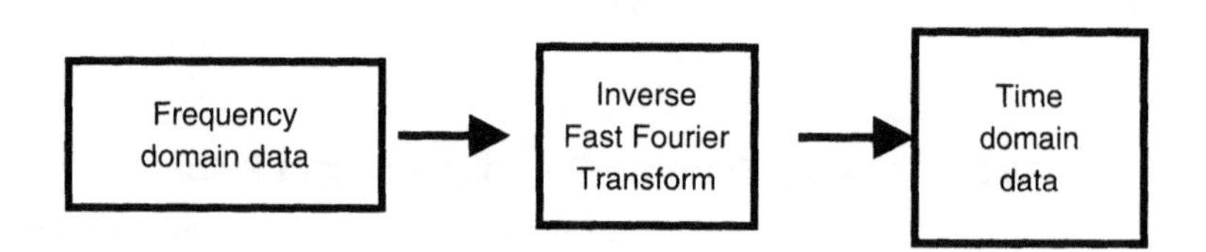

Figure 5.14

Operation of the Inverse FFT

Signal information in the frequency domain can be processed and then converted back into the time domain with an inverse Fourier transform.

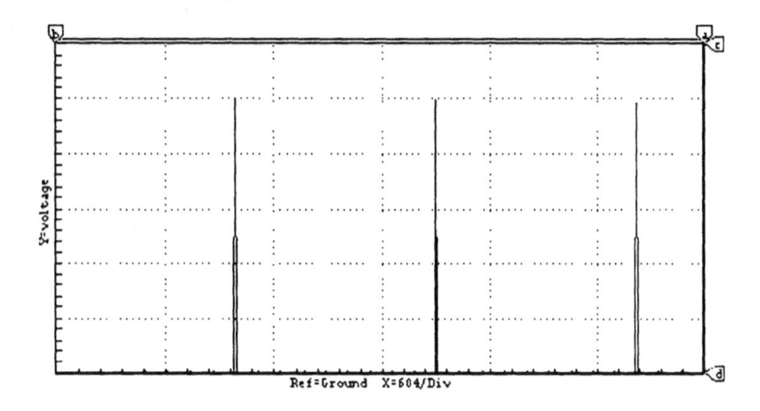

Figure 5.15

Type Complex Frequency Data Input

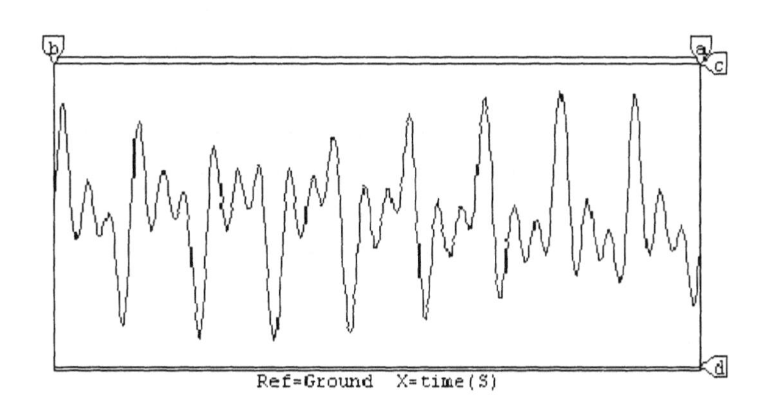

Figure 5.16

Time Domain Data Output

5.5 The Fast Fourier Transform (FFT)

The discrete Fourier transform as originally described by Fourier requires an exponential increase in the number of calculations as the sample size doubles. The DFT process executes a large number of redundant calculations, which also increase exponentially to the number of sample points. Performing a discrete Fourier transform on a sample set of 1,024 samples requires 1,048,567 complex multiplications. It was found (Runge, 1903; and Danielson & Lanczos, 1942) that it was possible to effectively fold a DFT into two halves. If the 1,024 samples are processed as two 512 sample sets, then each half takes only 262,144 multiplications each. The total combined calculations is therefore 524,288 multiplications, a savings of 50%.

If it's possible to fold a 1,024-point sample set into two 512-point sample sets, and still get the same answer, why can't each 512-point DFT be folded down into four 256-point sections for an even greater reduction? The process of folding the sample set and removing the redundant calculations is referred to as decimation. You can actually perform a DFT on a two-point sample set, so the idea behind the fast Fourier transform is to perform the calculations on a sample set that has been folded into binary multiples. This is why the FFT requires a sample size that is a power of two. A Fourier transform algorithm that breaks the sample set into sections of two samples each is called a Radix-2 implementation. On some processors, a Radix-2 algorithm can take longer to perform the decimation than it takes to do the complex multiplications. Some implementations of the FFT use a Radix-4 decimation, which breaks the sample set into sections of 4 samples each.

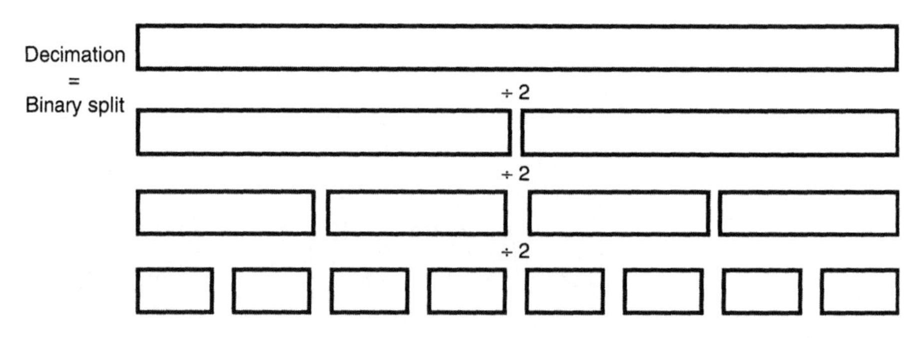

Figure 5.17

Binary Decimation

5.5.1 FFT Implementation

The implementation of a computer algorithm to calculate the Fourier transform by performing binary decimation was developed by James W. Tukey and J.W. Cooley of IBM Research. That algorithm has become known as the Fast Fourier Transform, or the FFT. A researcher at IBM, Richard Garwin, was studying solid helium and had a pressing need for Fourier techniques. Garwin contacted Tukey, who supplied him with a conceptual outline of how the Fourier transform might be evaluated on a computer. Garwin took the idea to J.W. Cooley to have the algorithm programmed.

"Garwin came to the computing center at IBM Research in Yorktown Heights to have the algorithm programmed. I was new at the computing center and was doing some of my own research. Since I was the only one with nothing important to do, they gave me this problem to work on.

It looked interesting, but I thought that what I was doing was more important. However, with a little prodding from Garwin, I got the problem out in my spare time and gave it to him. It was his problem, and I thought I would hear no more about it and went back to doing some real work."—*J.W. Cooley*[2]

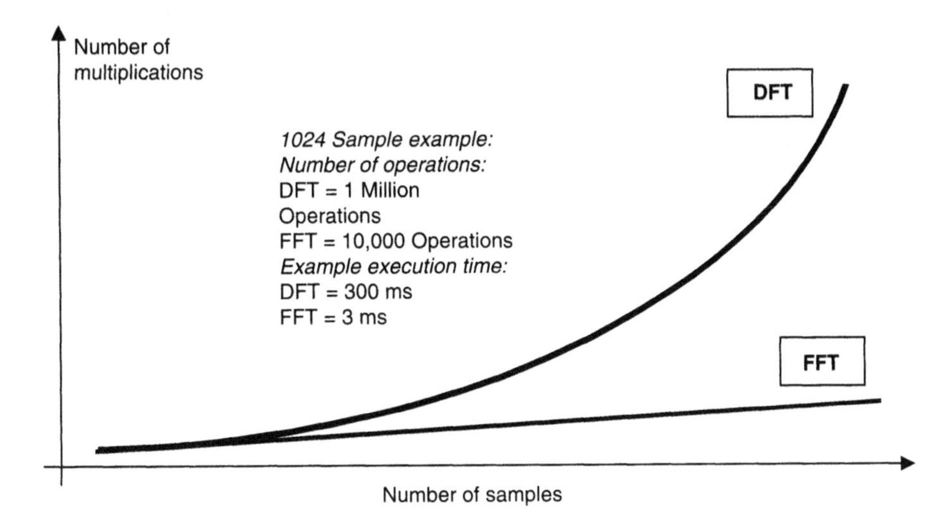

Figure 5.18

Comparison of the DFT and FFT Transforms

5.5.2 The Operation of the FFT

Except for some digital noise and the order of the coefficients, the results of the DFT and the FFT are the same. The process of multiplying the time domain data set times the sine and cosine is the same, but the FFT is more efficient. The FFT generates results for all of the spectral outputs up to one-half of the sample frequency. In cases where a single frequency spectrum is needed, the DFT can be faster. The output of the FFT generates data on increments of fbase (fbase × 1, fbase × 2, fbase × 3...).

sample frequency = 102400 Hz
sample size = 1024
fbase = 100 Hz

[2] (from *IEEE Transactions on Audio and Electronics* June 1969, quoted by Robert W. Ramirez, *The FFT, Fundamentals and Concepts*)

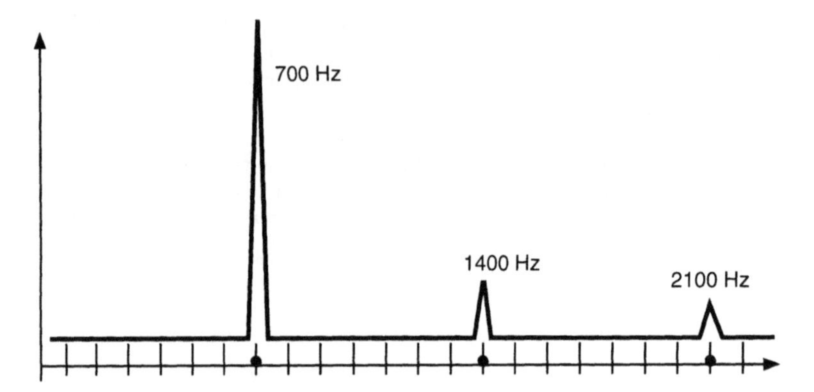

The DFT can be viewed as a tunable filter.

The FFT can be illustrated as a **filter bank**, arranged in increments of fbase.

fbase * 0
fbase * 1
fbase * 2
fbase * 3
fbase * 4
fbase * 5
fbase * 6
fbase * 7
...

Figure 5.19

The FFT as a Tunable Filter

5.6 Interpreting Frequency Domain Data

Now that we've got it, what can we do with it?

- Harmonic distortion tests
- Signal-to-noise ratio tests
- Frequency range tests
- Brick wall filters
- Mathematical over-sampling

5.6.1 fbase and the Frequency Bin

The output of the FFT is complex in type. Most test applications convert the complex data to magnitude information. Once the data is in magnitude format, the frequency domain data can be viewed as an array of energy values ranging from 0 Hz to fs/2.

The value in each element in the array represents the energy for the element location × fbase.

$$\text{cycles} \equiv \left(\frac{\text{fi}}{\text{fbase}}\right) = \text{frequency bin}$$

The action of the FFT is to sort the signal frequencies into a sequential set of frequency bins. To locate the bin for a specific frequency component, divide the frequency by the calculated fbase. The cycles value is identical to the frequency domain array element location for that frequency. In the time domain, the cycles value indicates the number of cycles in the data set. In the frequency domain, the cycles value indicates the signal frequency location, or the frequency bin location.

Example

sample frequency = 102400 Hz
sample size = 1024
fbase = 100 Hz

The frequency domain data set is an array of amplitude values, arranged in increments of fbase. To locate the frequency domain amplitude component of 600 Hz, divide the target frequency by the fbase value.

$$\text{cycles} \equiv \left(\frac{\text{fi}}{\text{fbase}}\right) = \frac{600\ \text{Hz}}{100\ \text{Hz}} = 6$$

The frequency domain data set contains the amplitude for the 600-Hz component at array element 6.

5.6.2 Interpreting Frequency Domain Magnitude Values

Now that you've got frequency domain data organized into fbase increments, the next question is what happens to the amplitude information? The action of the polar conversion on the type complex frequency domain data produces some goofy numbers that do not intuitively correlate to the original time domain signal amplitude. For example, let's say the 600-Hz signal described in the previous example had a peak amplitude of 693 mv in the time domain. After the FFT and the polar conversion, the frequency domain magnitude value would be 1419.24. All those complex multiplications wind up doing some weird things to the amplitude value, but it is fairly straightforward to "reverse engineer" the results.

If you are making relative measurements with the frequency domain data, such as noise or distortion, you may not need to map the frequency domain magnitude values back to the original signal amplitude. Even though the raw numbers may look a little strange, all of the fbin magnitude values keep the correct ratios. If you do want to map the magnitude values of the FFT back to the time domain peak voltage, here's what you do:

For polar data:

If the FFT complex data set was processed with a polar conversion to produce

magnitude data

1. Divide the magnitude value by the time domain sample size.
2. Divide the results by 2 (except for the DC component).

$$\text{peak} = \frac{\left(\dfrac{\text{magnitude}}{\text{samples}}\right)}{2}$$

If the FFT complex data set was processed with a complex vector magnitude (cvmags) conversion to produce magnitude data,

1. Get the square root of the magnitude value.
2. Divide the results by the time domain sample size.
3. Divide the results by 2 (except for the DC component).

$$\text{peak} = \frac{\left(\dfrac{\sqrt{\text{magnitude}}}{\text{samples}}\right)}{2}$$

5.6.3 Nyquist and Shannon

Recall from the Nyquist and Shannon theorems that

- The sample rate must be greater than twice the signal frequency.
- The sample set must have more than two samples per cycle of the signal.

What are the implications in the frequency domain? Among other things, it means that the data set is valid from DC to one-half of the sample frequency. If a signal component is greater than one-half of the sample frequency, that means there were not more than two samples per signal cycle. As it turns out, the mathematical process involved with the complex Fourier transform actually generates a mirror image of the data set, which is arranged as negative frequency. This data is redundant and is ignored.

The number of data elements in the frequency domain data set can be found by dividing the Nyquist frequency (fs/2) by fbase. That is, find the frequency bin value for Nyquist.

$$\text{frequency\_bin} \equiv \left(\frac{\text{Nyquist}}{\text{fbase}}\right)$$

The result is that the number of data points in the frequency domain is one-half the number of data points in the time domain. Using our previous example,

sample frequency = 102400 Hz
sample size = 1024
fbase = 100 Hz

In order to capture at least two samples per cycle, the signal frequency can be no greater than one-half of the sample frequency. Because the sample frequency is 102.4 kHz, the signal frequency bandwidth extends to 51.2 kHz. The frequency domain data set extends from DC to Nyquist, in increments of fbase. In other words, the fbase value is the "step size" of the frequency domain array. The number of frequency fbins is always one-half the time domain sample size.

$$\text{frequency\_bin} \equiv \left(\frac{\text{Nyquist}}{\text{fbase}}\right) = \frac{51.2\ \text{kHz}}{100\ \text{Hz}} = 512$$

5.6.4 Harmonic Distortion Tests Using the FFT

The stimulus data for a harmonic distortion test is a single-tone sine wave, known as the fundamental frequency. Errors in the output of the device that occur at multiples of the fundamental frequency are referred to as harmonics. To test for harmonic distortion, the frequency domain data is analyzed by first measuring the amplitude of input signal frequency, that becomes the reference point for the harmonic content ratio. The amplitude for the frequencies that are integer multiples of the signal frequency are measured and summed, and then the results are calculated as a percentage or a dB ratio.

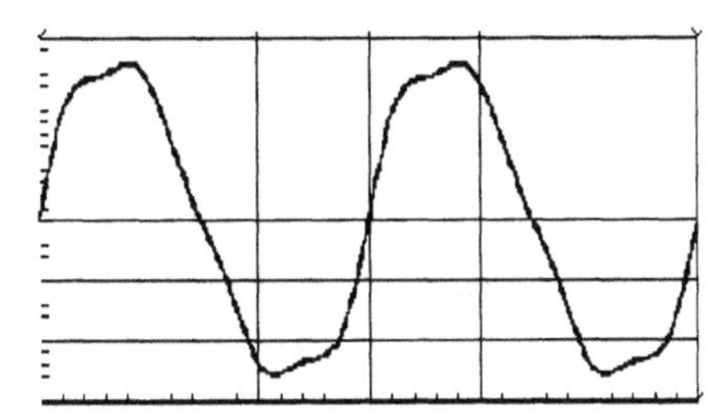

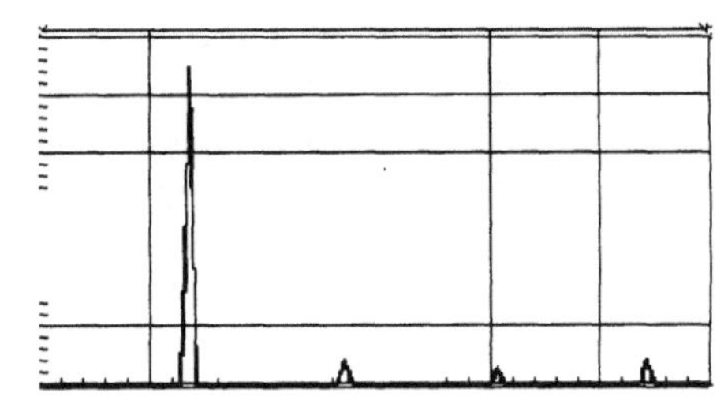

Figure 5.20

Distortion in the Time Domain and the Frequency Domain

Step One: The DUT output is a distorted 5-kHz sine wave. The signal is captured at a 32-kHz clock rate, with 256 samples. The fbase is therefore 125 Hz.

Step Two: The captured signal is processed with an FFT. The frequency domain data set contains the amplitude for frequencies from DC to 16 kHz (fs/2), in steps of 125 Hz.

Step Three: The fundamental signal, at 5 kHz, is located as frequency bin = fi/fbase, 5000 Hz/125 Hz = 40. The fundamental signal energy, located in the frequency domain array element number 40, is 0.97 volts.

Step Four: The amplitude of signal frequencies found at integer multiples of the fundamental are located and measured as follows:

Second Harmonic: Frequency = 10 kHz. Array Element = 80. Amplitude = 42 mV
Third Harmonic: Frequency = 15 kHz. Array Element = 120. Amplitude = 15 mV

Step Five: The total amount of distortion error is calculated as the algebraic sum of the harmonics.

$$\text{Harmonic\_Error} = \sqrt{\text{2ndHarm}^2\text{\_3rdHarm}^2} \quad \sqrt{42\ \text{mV}^2 + 15\ \text{mV}^2} = 44.6\ \text{mV}$$

Step Six: Calculate the harmonic error as a percentage of the fundamental signal level.

$$\text{THD\%} = \frac{\text{Harmonics}}{\text{Fundamental}} \times 100 = \frac{44.6\ \text{mV}}{0.97\ \text{V}} \times 100 = 4.59\%$$

5.6.5 Signal-to-Noise Tests Using the FFT

The stimulus data for a signal-to-noise ratio test is a single-tone sine wave fundamental frequency. Noise is usually defined as spurious signal energy in the output signal of the device, which occurs at non-harmonic intervals of the fundamental frequency. Testing for the SNR processes the result of the FFT to remove the energy components due to the DC value, the fundamental signal energy, and optionally the harmonic energy. The remaining data is summed together to derive a total noise value across a defined frequency span as the square root sum of the squares.

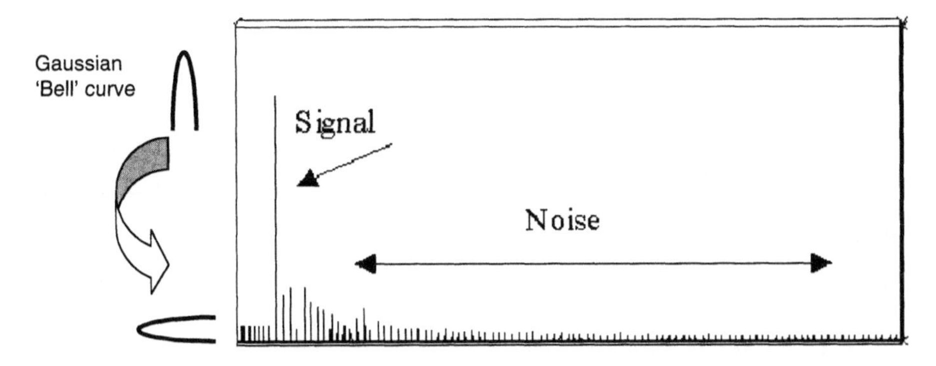

Figure 5.21

Gaussian Distribution of Noise Amplitude

When we talk about Gaussian distribution when measuring noise, we're referring to the statistical variations of amplitude levels of the noise components across the spectrum. If you graph a histogram of random noise amplitude levels, it should look like a bell curve. Because noise measurements are referenced to a Gaussian distribution, the SNR calculation must include a statistically valid number of noise amplitude sample points. Assuming a normal distribution, a sample set of 100 samples or more will follow the rules of a large statistical population. Because the FFT frequency domain data set is one-half the size of the time domain data set, and the FFT requires a power of two, we can conclude the following:

1. The statistical sample size must be greater than 100 samples, and also be a power of two. (A power of two that is greater than 100 is 128.)
2. If the frequency domain data set is 128 samples, then the time domain data set size must be 256 samples.

With a 128-element frequency domain data set, the signal-to-noise test first records the amplitude of the signal. The frequency domain values corresponding to DC and the signal harmonics are set to zero, and the remaining frequency domain values are processed to extract the square root sum of squares.

$$\text{Noise} = \sqrt{\text{n}1^2 + \text{n}2^2 + \text{n}3^2 + \cdots + \text{n}^2}$$

where "n" is the amplitude of a noise component in the frequency domain.

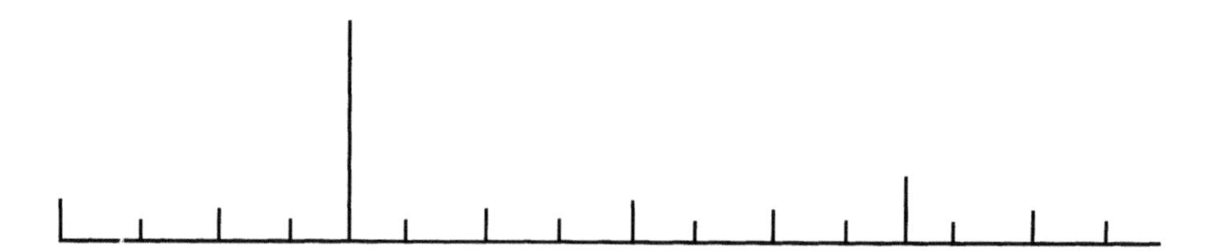

Figure 5.22

Noise Distribution in the Frequency Domain

1. Measure the signal amplitude.

DC	n1	n2	n3	S	n4	n5	n6	2H	n7	n8	n9	3H	n10	n11	n12

Figure 5.23

Measure the Signal Amplitude

2. Null out the DC, signal, and harmonics.

0	n1	n2	n3	0	n4	n5	n6	0	n7	n8	n9	0	n10	n11	n12

Figure 5.24

Zero-out the Fundamental and the Harmonic Energy

3. Extract the algebraic sum of the noise components.

$$\text{Noise} = \sqrt{\text{n1}^2 + \text{n2}^2 + \text{n3}^2 + \cdots + \text{n}^2}$$

The ratio of the signal amplitude to the noise level is expressed in terms of decibels (dB).

$$\text{SNR} = -20 \times \log\left(\frac{\text{Signal}}{\text{Noise}}\right)$$

Some applications make a distinction between signal energy that occurs at integer multiples (harmonics) of the input signal frequency and energy at non-harmonic intervals. A signal-to-noise measurement that includes the distortion component is referred to as S/N + D (signal to noise plus distortion) or SNDR. The S/N + D test is sometimes listed as a "sinad" test, which is pronounced SINE-ADD. The test process for the S/N + D test is identical to the SNR test, except that the harmonic frequency domain components are processed along with the noise components.

5.6.6 Brick Wall Filters

Processing signal data in the frequency domain allows the signal to be digitally filtered. One of the advantages of processing information in the frequency domain is the precise delineation between the signal component and spurious components. If the spurious components can be identified, they can also be removed.

A "brick wall" filter gets its name from the extremely sharp cut-off frequency. The digitized signal is converted to frequency domain data by the FFT. The output of the FFT is processed to remove all of the energy components above the desired frequency limit. After the unwanted energy has been removed, the time domain signal can be reconstructed by means of a reverse FFT. A reverse FFT transforms the data from the modified frequency domain back into time domain data.

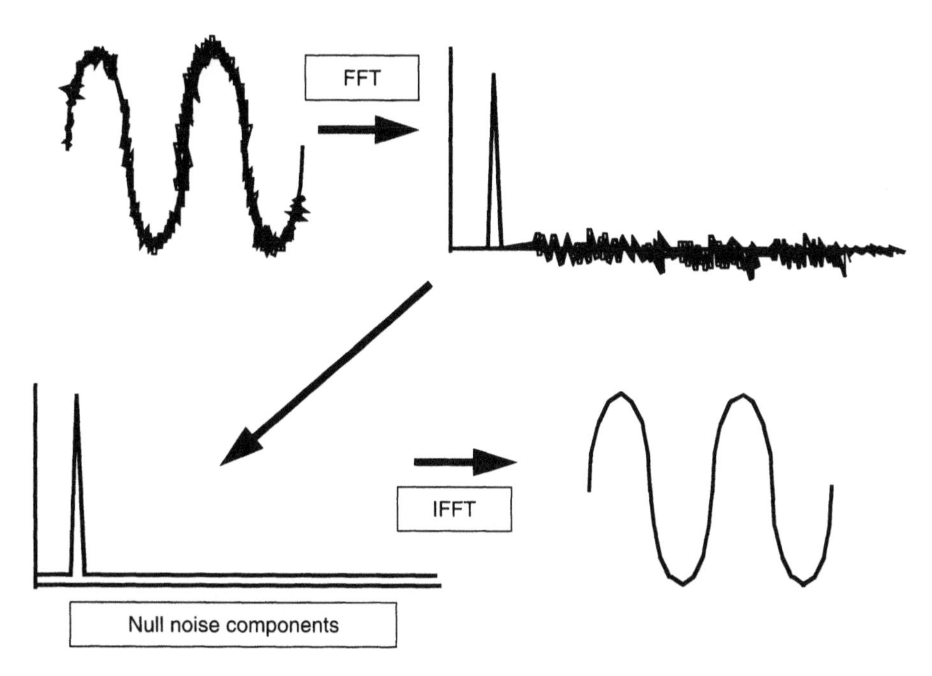

Figure 5.25

The Brick Wall Filter

5.6.7 Mathematical Over-Sampling

Mathematical over-sampling is another way of processing the digitized signal in the frequency domain. In applications where it is not possible to capture the signal with a sufficient number of samples for each cycle, it is sometimes useful to interpolate between sample points. Mathematical over-sampling does not create any new data, it simply reconstructs a signal by estimating data between existing points.

To execute mathematical over-sampling, the digitized data is converted to the frequency domain. Recall that the span of data in the output of a FFT is equal to one-half of the sample frequency. Increasing the sample frequency for a given window duration has the effect of increasing the number of samples, and extending the FFT data set span. Mathematical over-sampling simulates a higher sample rate by extending the frequency domain data set, as if the sample frequency had been increased.

The frequency domain data is processed by appending to the frequency domain data set, which in effect extends the bandwidth. The appended frequency domain data is then processed with a reverse FFT. Instead of the frequency domain data being limited to one-half of the sample frequency, the new data is reconstructed based on the artificially created frequency range. The effect of the reconstruction is a sine x/x interpolation between the original data points.

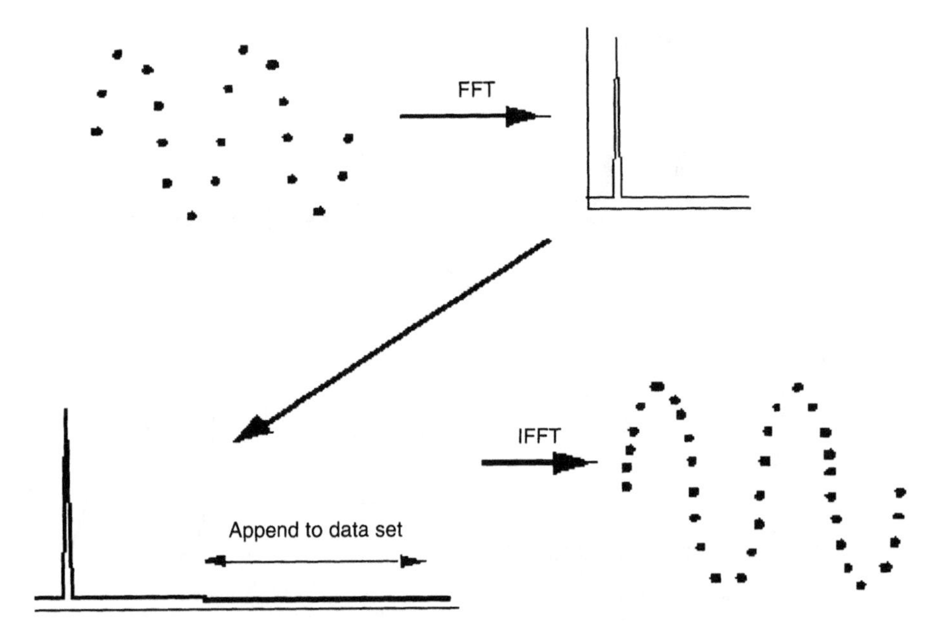

Figure 5.26

Mathematical Over Sampling

5.7 Minimizing Frequency Domain Anomalies

Anomalies in the time domain data set can result in erroneous data in the frequency domain. Once frequency domain errors are present, it often is very difficult to separate valid data from spurious information. Some of the time domain characteristics that can degrade the result of the FFT are

- Non-periodic Sample Sets
- Leakage
- Settling Time
- DC Offset
- Noise
- Aliasing

Small errors in the time domain can create large errors in the frequency domain. However, by understanding and anticipating possible sources of error, frequency domain anomalies can be minimized.

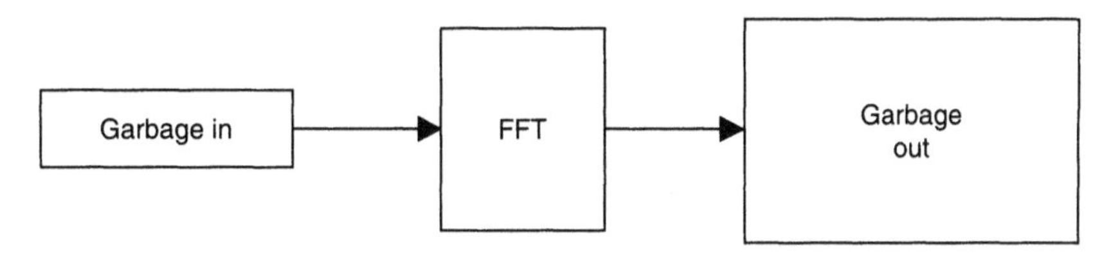

Figure 5.27

Anomalies in the Frequency Domain

5.7.1 Effects of the Time Domain Window

The process of generating a finite sample set establishes a beginning point and an ending point, with a finite number of discontinuities (on and off). This meets the Dirichlet requirements for a sample set, which states that a time domain sample set is valid if it has a "finite number of minima and maxima, and a finite number of discontinuities." The FFT will transform the time domain data of the sample window itself into frequency domain data. In other words, the Fourier transform cannot tell the difference between the signal and the window in the time domain data set. Both the signal information and the window information are processed together.

Without additional processing, the action of turning the digitizer on and off within a period of time generates a rectangular shape in the time domain. When transformed into the frequency domain, a rectangular wave shape generates a sine (X) over X frequency distribution. Referring to the earlier illustration of the FFT as a bank of filters, we can view this frequency domain representation of a rectangular window as the actual "shape" of the filter response.

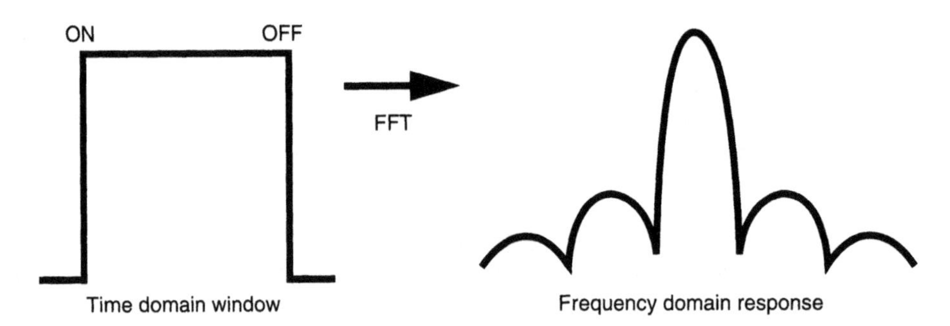

Figure 5.28

Energy Distribution of the FFT Filter

Early in the chapter, the Fourier transform was described as a filter, and the illustration showed a very flat and steep bandpass filter response. Actually, it's not really a rectangular shape at all.

Instead of a filter response that looks like Fig. 5.29(a), it's really more like Fig. 5.29(b).

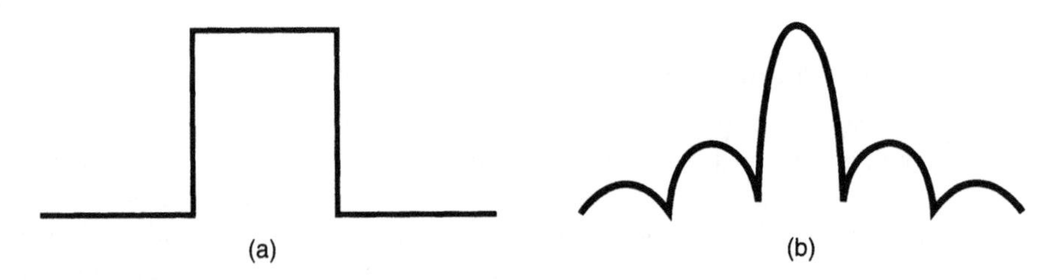

Figure 5.29

Sine (X) over X Filter Response

(Yeah, I know it's a lot different, but I figured we had to start somewhere.)

1. The process of generating a finite sample set in the time domain generates data points that are processed by the FFT.

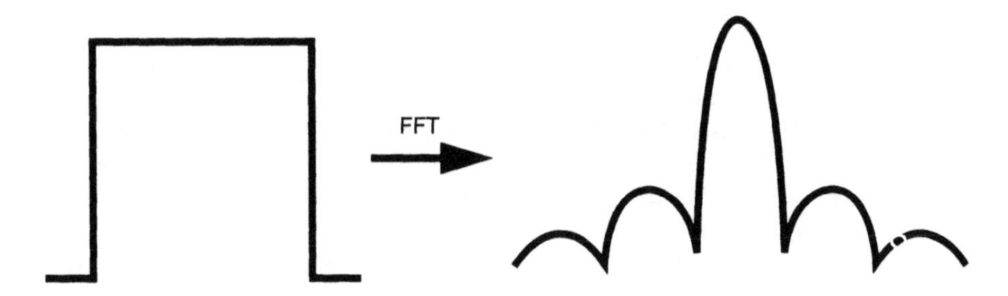

Figure 5.30

Effects of the Time Window in the Frequency Domain

2. If it were possible to capture an infinite number of cycles, the time window information would not be a factor. Without the time window data, the FFT would convert the signal into a single line spectra.

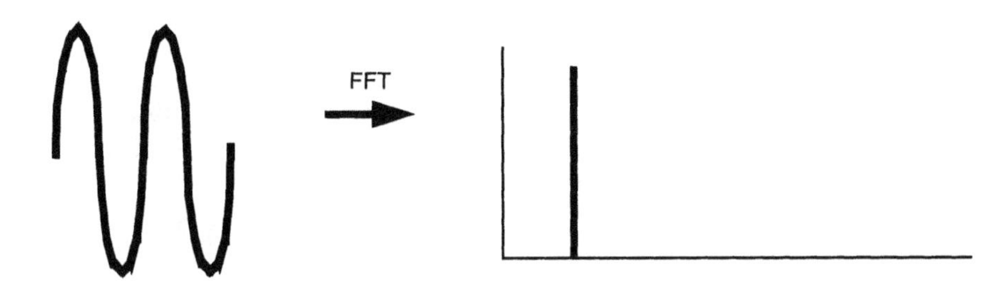

Figure 5.31

Single Line Spectra

3. When the sample window contains a signal, the signal is processed as time domain data that is converted to frequency domain data by the FFT. The time domain data of the signal and the window data are rolled together by the action of the FFT. The two data sets, signal and window, are effectively multiplied together.

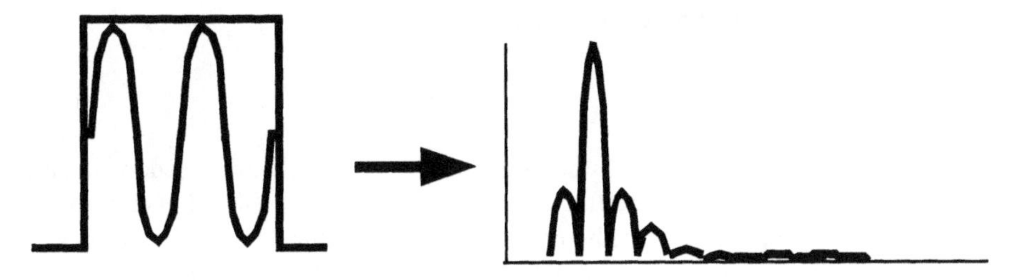

Figure 5.32

Convolution of Window and Signal Data

5.7.2 The Effects of Convolution

My friend Rich Wilhelm has a great illustration of the effects of convolution: *The Fly*. Now, if you did not have a dissipated youth like I did, allow me to outline the plot of this classic film. Dr. Seth Brundel is a scientist who is trying to build a transporter machine, just like they had on *Star Trek*. Eventually, Brundel tries the transporter with himself as the cargo, before it is completely "de-bugged." (There I go again. Sorry!). He pushes the button to get transported, not knowing there is a fly in the transporter room with him. Alas, Dr. Brundel and the fly are merged together in the transporting process. That, dear reader, is convolution. He's no longer Seth, it's no longer a fly—but both have characteristics of both data sets.

The result of windowed data in the frequency domain is that the data has the amplitude and the frequency location of the signal, and the sine (X) over X shape of the window. The action of the FFT is to convolve a complex frequency domain data set of a single rectangular pulse (sine (X)/X impulse) with a complex data set containing positive frequency values. The output has features of both input data sets—the signal, and the window.

Fortunately, the FFT acts like a filter bank at fbase increments. The sine (X) over X energy from the window convolution is always at ZERO at fbase increments. As long as our data stays on fbase boundaries, the effects of the window energy convolution will not be a problem.

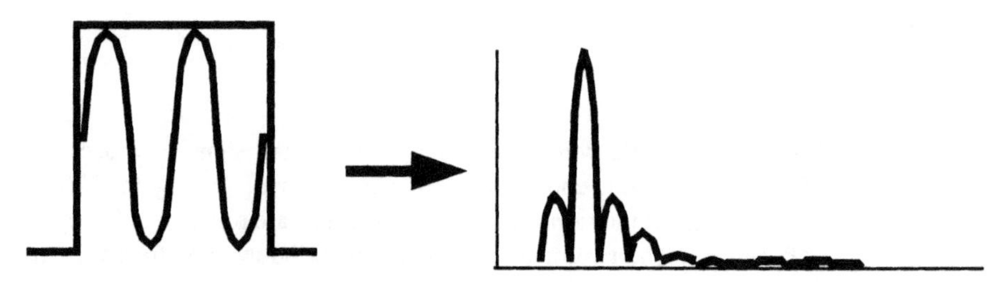

Figure 5.33

Effects of Convolution in the Frequency Domain

However, signal data that is not periodic will result in frequency domain energy that is not at multiples of the frequency resolution (fbase). Signal energy that is not at fbase increments is not filtered by the action of the FFT, and instead is summed with the side-band energy from the window data. The signal data leaks into adjacent frequency bins.

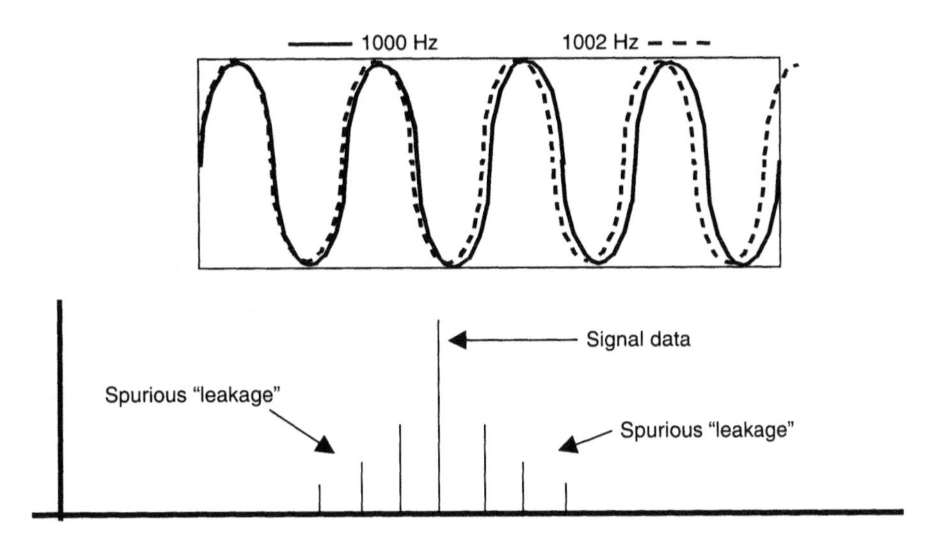

Figure 5.34

Leakage and Smearing

5.7.3 Frequency Domain Leakage

Leakage is another way of describing non-periodic data in the time domain that is not aligned with the base frequency. Because there is signal data that does not have an integer relationship to the window duration, the signal energy "leaks" into adjacent locations. If the signal is not "lined up" with the center of the filter, then some of the signal amplitude is rejected and falls into an adjacent frequency bin.

A. Filter frequency and signal frequency are the same. The signal is at the center of the filter response. All of the signal energy corresponds to the specific frequency bin.

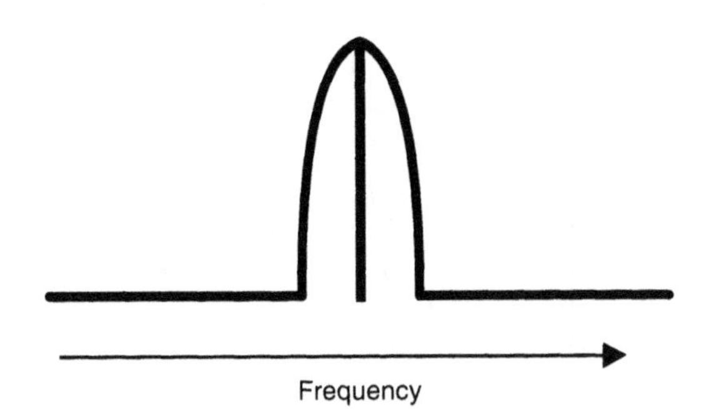

Figure 5.35

Signal Data Aligned with Frequency Domain "fbase" filter

B. The signal frequency is slightly higher than the center of the filter. Most of the signal energy will fall into the specific fbin, but not all. The signal energy that is rejected by the filter will "leak" into adjacent frequency bins.

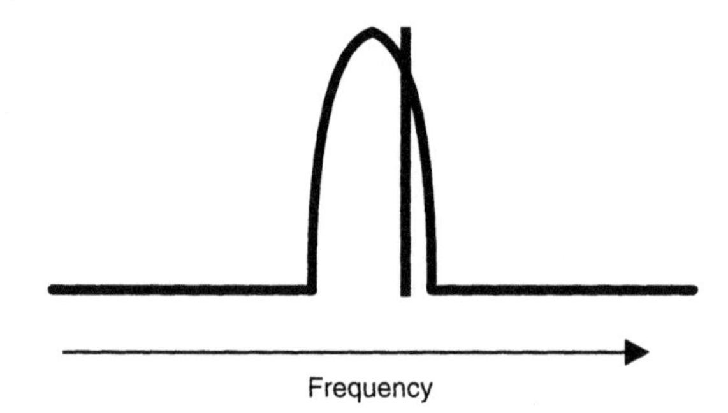

Figure 5.36

Signal Data Misaligned with Frequency Domain "fbase" Filter

The result of leakage error is that power is taken from frequency components existing in the time domain signal and transferred to frequency components that do not exist in the time domain signal.

A common source of leakage is radiated signal from the 50/60-Hz power frequency. This illustration shows a 250-Hz signal, superimposed with a spurious 60-Hz line frequency component. Even though the sample set is periodic for the signal frequency, the window is not periodic for the 60-Hz component. As a result, the frequency domain data set generated by the FFT will contain spurious low-frequency information.

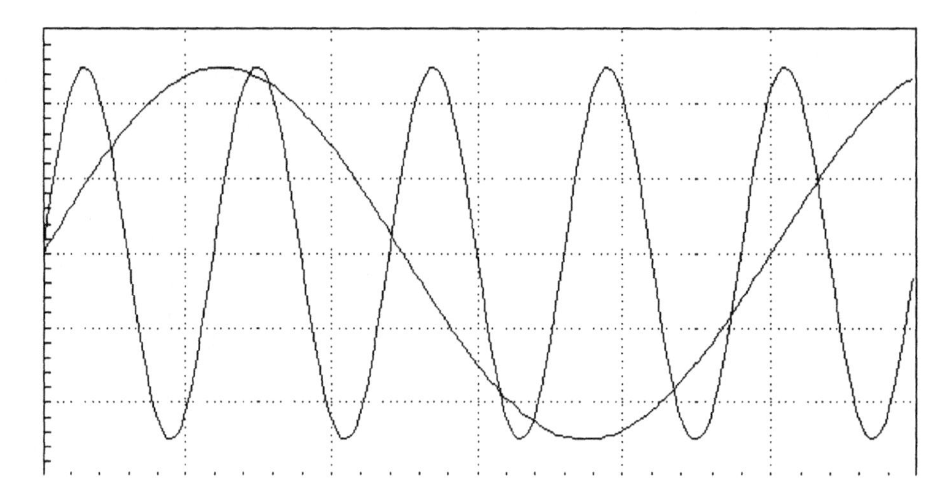

Figure 5.37

Non-periodic Time Domain Data Causes FFT Leakage

Solutions to Frequency Leakage Error

1. Make sure the fbase ratio is an exact integer sub-multiple of the signal frequency.
2. If the introduction of spurious signal information cannot be avoided, choose an fbase ratio that can accommodate both the signal and the spurious frequency.
3. Keeping in mind that the frequency domain resolution is an inverse function of test time, it may not be practical to choose a sufficiently small fbase. In that case, a technique known as windowing may provide a solution.

5.7.4 Windowing Time Domain Data

In applications where the time domain data cannot easily be captured as a periodic data set, it is often helpful to process the time domain data to modify the signal amplitude. A rectangular window has the same amplitude scalar (1.0) for every sample. Other window shapes can be applied that modulate the amplitude of the captured time domain signal data set. By modulating the time domain data with a variable scalar, the effects of the discontinuities are minimized. These modulation shapes are called windows.

By minimizing the signal amplitude at the end-points of the capture window, the spurious effects on non-periodic data are minimized. However, the effects of the signal attenuation must be accounted for when the modulated time domain data is converted to frequency domain data. Some common window algorithms are Hanning, Hamming, and Blackman. (My guess is that Hanning and Hamming must have sat next to each other at MIT, and decided to work on their thesis together.) Each algorithm has a unique modulation shape, and corresponding effects in the frequency domain. Window algorithms have the effect of reducing the effects of a non-periodic sample set, at the cost of reduced information and lower precision in the frequency domain.

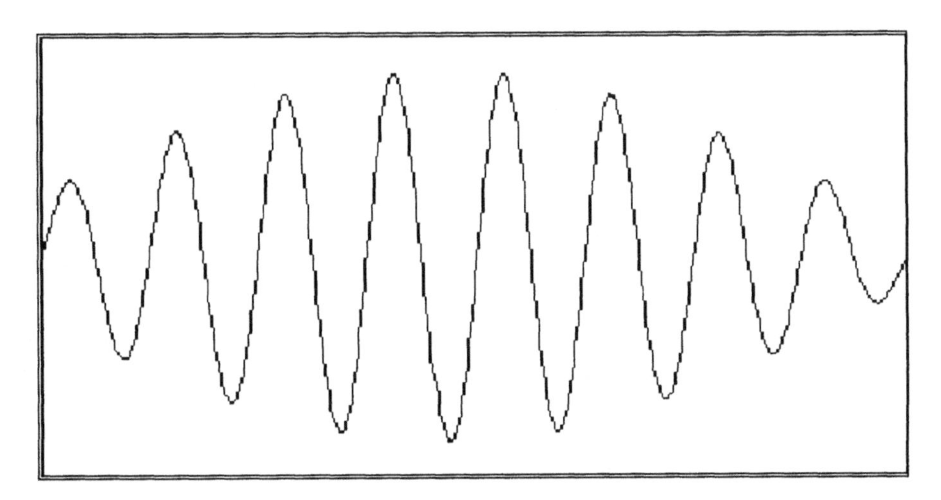

Figure 5.38

Windowing the Time Domain Data

5.8 Signal Aliasing

Nyquist's Theorem states that properly representing a waveform requires a sample rate of at least twice the signal frequency. Another way of applying Nyquist's theorem is to state that only sampled frequencies that occur below fs/2 can be properly processed. What happens to signals that are above the Nyquist frequency? The effect of aliasing is that signal frequencies above the Nyquist point (fs/2) are *folded back* to appear as lower frequency signals. It's not really the signal, but it sure looks like it. The alias is a ghost.

Example

If the digitizer is sampling at a rate of 16000 Hz, the Nyquist frequency is 8000 Hz. If the captured signal is at 10 Khz, the signal frequency data will be at 2 kHz above Nyquist. The alias, or "ghost" signal will appear as data 2 kHz *below* Nyquist.

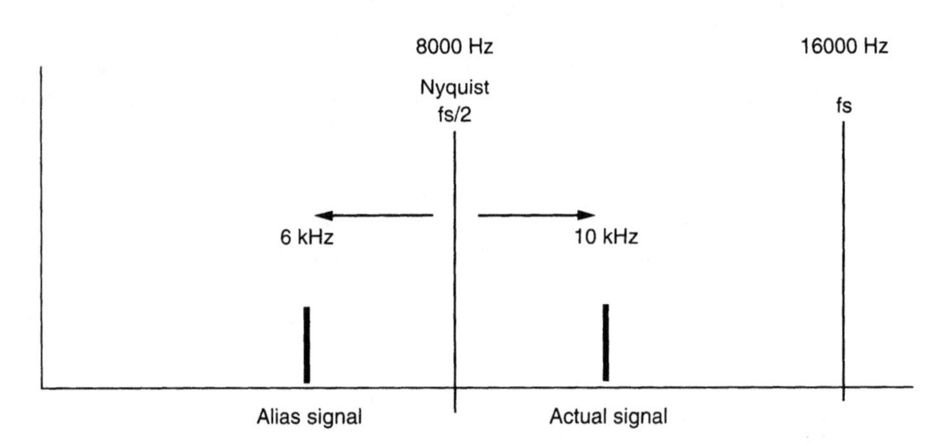

Figure 5.39

Signal Aliasing in the Frequency Domain

In the analog capture hardware, the anti-alias filter circuit is usually programmed to reject signals that are above the Nyquist point. Any spurious signal information that could cause an alias is attenuated by the low-pass function of the anti-alias filter. Signal frequencies that are above the sample frequency (fs) are "folded back" a second time, and appear as alias signals referenced to 0 Hz.

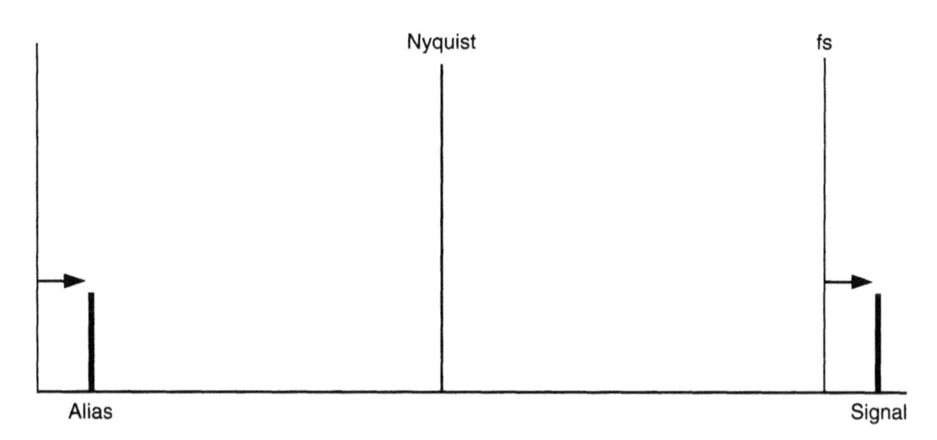

Figure 5.40

Under-sampling Measures the Signal Alias

5.8.1 Under-Sampling Techniques

In applications that have a signal frequency component above the Nyquist limit of the digitizer, the alias effect can be used to effectively extend the digitizer range. Let's say you've got a signal with a frequency of 1.2 MHz, but your digitizer has a maximum clock rate of 1.0 MHz. If you know where to look within the spectrum data set, you can indirectly measure the signal by measuring the alias.

Remember that the alias is a ghost? Well, it's a living, breathing, Technicolor ghost, because it is an exact duplicate of the original signal. As long as the digitizer hardware did not attenuate the signal, the amplitude of the alias is exactly the same as the amplitude of the original signal. Going back to the example, if the digitizer is clocking at 1.0 MHz, then the Nyquist point is 500 kHz. Because 1.2 MHz is 200 kHz above the sampling frequency (fs), the signal alias will show up at 200 kHz.

By carefully choosing a sample frequency, the application can place the Nyquist point at any value. As a result, the alias frequency of the high-frequency component can be precisely placed anywhere in the frequency spectrum. The amplitude of the alias will be exactly the same as the amplitude of the original signal, limited only by the bandwidth of the digitizer. It is therefore the bandwidth, not the maximum sample rate, that is the frequency limit of the digitizer.

Chapter Review Questions

1. What is "frequency domain data"? How is it different from "time domain data"?

2. What type of test is frequency domain data used for?

3. How is a "brick wall" filter implemented with an FFT?

4. Why does an FFT need a power-of-two (2^n) sample size?

CHAPTER 6

DSP BASED TESTING

'Contrariwise,' continued Tweedledee, 'if it was so, it might be, and if it were so, it would be; but as it isn't, it ain't. That's logic!

—Lewis Carroll, *Alice in Wonderland*

6.1 Introduction

This is the fun part! DSP lets you slice, dice, mix, match, mangle, and measure the waveform data any which way. The use of digital signal processing for test engineering applications provides tremendous advantages over conventional analog instrumentation. The DSP can be used to analyze the captured signal with better accuracy and at higher speeds. Measurements that may have been difficult or impossible using analog-based instruments are relatively straightforward using DSP techniques.

The main types of processing that can be performed on a DSP include array management, vector and scalar math, format conversions, and complex transforms. Back when I was a young man, these gadgets used to be called an array processor. That's a good name, because a DSP involves the manipulation of signal data as an array.

Most computers are designed to process data as single elements. A DSP system, in contrast, performs calculations and manipulation on data arrays.

The array to be processed may represent an analog signal in either the time or frequency domain, or may represent derived information. The DSP in a test system is sometimes still referred to as an *array processor*. The DSP in a test system can be programmed to perform measurements as a virtual instrument. The DSP system does not control measurement hardware, it replaces it.

DSP Processor algorithm—(conceptual illustration)

```
float x[5] = {1.0,2.7,1.5,3.9,4.3};
float y[5] = {2.0,4.1,1.7,3.2,2.2};     <<declaration section
float z[5] = {0.0,0.0,0.0,0.0,0.0};

add
     in1 = x
     in2 = y                      <<algorithm section
     output = z
     iteration = 5;
```

6.2 DSP Algorithm Structure

DSP algorithm statements look kind of hairy at first, but they're easy to unravel once you know how. The statements define *what* operation to perform, *which* array to perform it on, and *where* to put the result. Because DSP algorithms process arrays, the algorithm statement must also declare the iteration size. Reviewing the previous example,

```
add                  <<<WHAT to do
in1 = x              <<<on WHICH arrays
in2 = y
output = z           <<<WHERE to put the result
iteration = 5;       <<<HOW MANY iterations
```

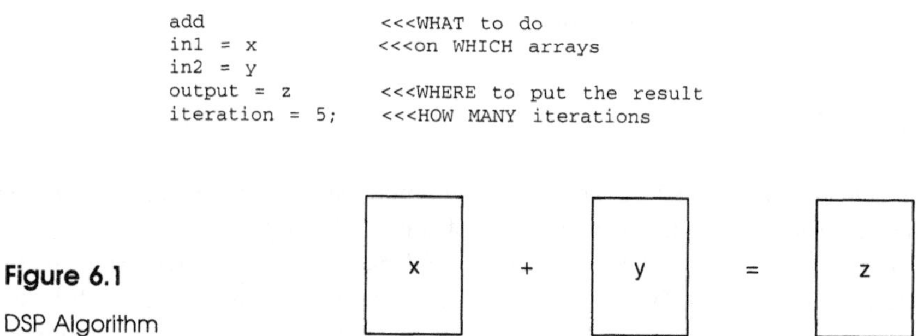

Figure 6.1

DSP Algorithm Diagram

6.2.1 Copying Arrays

Let's look at a few examples. A DSP copy algorithm creates a new array with identical contents. The copy algorithm can also copy a specified section from a source array into the destination array. The command requires that the destination array have the same data type and adequate size. The command also specifies the offset, or start point, in both the source array and the destination array.

```
/* copy the contents of array A to array C pseudo code example */
        copy_array
           from: dsp array A
                 input stride = 1
                 offset = 0
           to: dsp array C
               output stride = 1
               offset = 0
           iterations = 10;
```

6.2.2 Offset Parameter

By changing the offset, the algorithms can be programmed to operate on specific sections of an array. This can come in handy when you want to pull out a particular section of the signal.

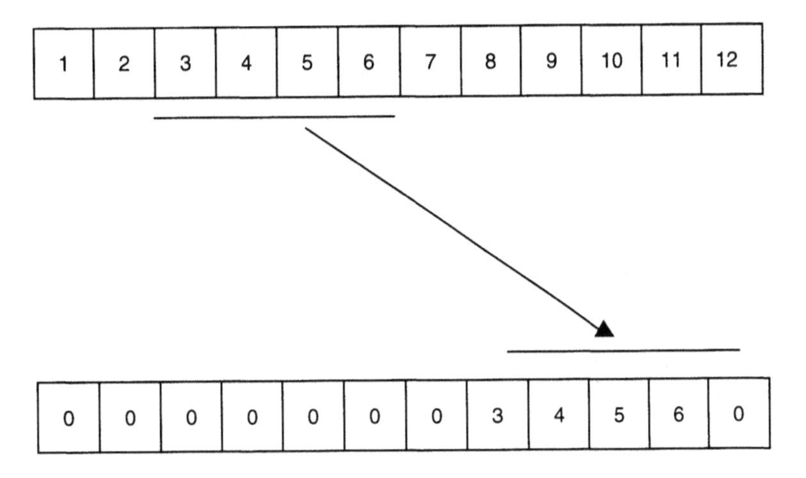

Figure 6.2

The Offset Parameter Changes the Start Point

```
copy_array
   from: dsp array A
         input stride = 1
         offset = 2
   to: dsp array C
       output stride = 1
       offset = 7
   iterations = 4;
```

6.2.3 Step Size Parameter

When I was a kid, I'd spend the summer with my grandfather. Grandpa had a little farm, and it was my job to help take care of the cow. Now, I don't know if you have ever walked across a cow pasture, but it is quite an experience. The trick is to change the size of your stride as required to avoid getting cow poop on your shoes. Well, you can also change the size of the stride with your DSP. Most DSP algorithms have a default step size of 1. That is, each data element in the input arrays will be processed sequentially, and each data element in the output will be written sequentially and incrementally. Some DSP algorithms allow the step size to be modified. Instead of performing the specified algorithm on every element in order, striding allows every *n*th element to be processed. Striding is an optional parameter for specific DSP algorithms. Algorithms may allow the input arrays and output arrays to be processed according to specific stride values. Data striding modifies the algorithm iteration count.

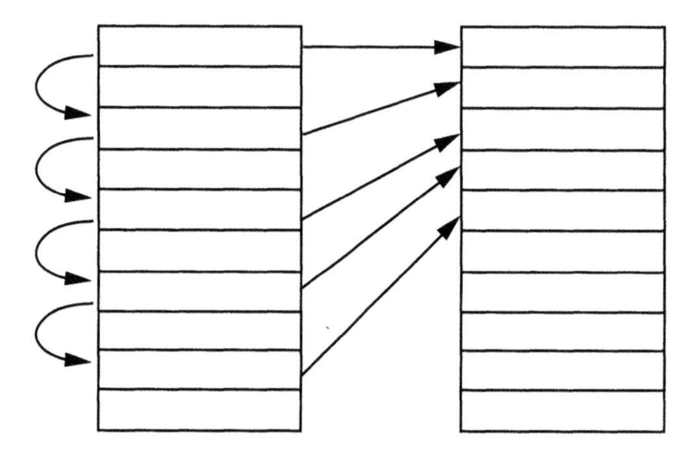

Figure 6.3

The Stride Parameter Changes the Step Size

```
/* copy every other element of array A to array C */
copy_array
       from: dsp array A
             input stride = 2
             offset = 0
       to: dsp array C
           output stride = 1
           offset = 0
       iterations = 5;
```

6.2.4 Application Example

DSP algorithms are available for setting a specified section of elements within an array to a specific value. The clear algorithm sets the specified elements to zero. The fill algorithm sets the specified elements to a defined value.

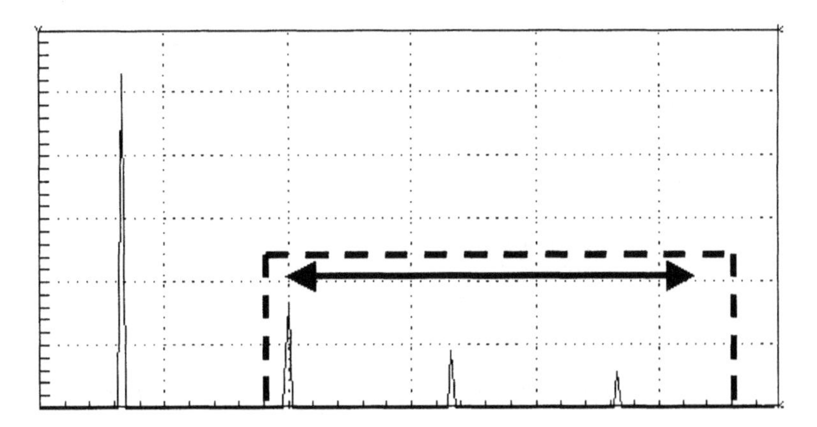

Figure 6.4

Vector Clear Example

The clear algorithm, for example, could be used as part of a brick wall filter sequence. The signal would be converted to the frequency domain via an FFT, and then the spurious signal data could be removed with the clear algorithm.

```
complex array fft_results[256];
vector_clear /* pseudo code example */
       input = fft_results
       offset = 128
       iteration = 128;
```

6.3 Math Operations

DSP systems use arrays as the data elements for math functions. The two types of math operations are vector and scalar. Vector functions perform the operation across each element of the input arrays, whereas a scalar operation uses a single value as the operand.

6.3.1 Vector Operations

Vector math algorithms work like a parallel function. Vector addition, for example, adds the first element of the input1 array to the first element of the input2 array, and puts the result into the first element of the output1 array. Then the algorithm repeats the process with the second element, then the third, and so on.

Vector Add

```
vector_add
     input1 = array_a
     input2 = array_b
     iteration = 10
     output = array_c;
```

This algorithm would take the first element of array_a, add it to the first element of array_b, and place the results in the first element of array_c. The process would be repeated for each element.

Vector Multiply

```
vector_multiply
     input1 = array_a
     input2 = array_b
     iteration = 10
     output = array_c;
```

This algorithm would select the first element of array_a, multiply it with the first element of array_b, and place the results in the first element of array_c. The process would be repeated for each element.

Vector addition

Array A	+	Array B	=	Array C
1		3		4
2		4		6
3		7		10
4		2		6
5		5		10
6		8		14
7		3		10
8		2		9
9		6		15
10		2		12

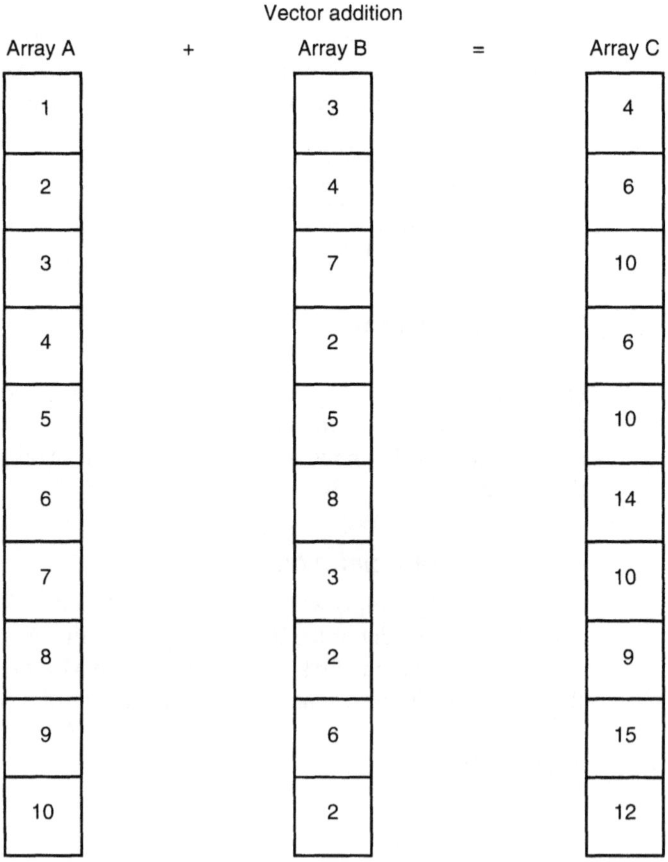

Figure 6.5
Vector Addition

Vector multiplication

Array A	+	Array B	=	Array C
1		3		3
2		4		8
3		7		21
4		2		8
5		5		25
6		8		48
7		3		21
8		2		16
9		6		54
10		2		20

Figure 6.6

Vector Multiplication

6.3.2 Scalar Operations

Scalar functions perform the operation on one input array by using a single element from a second array.

```
scalar_add /* pseudo code example */
     input1 = array_a
     scalar = array_b
     scalar_offset = 0
     iteration = 10
     output = array_c;
```

This algorithm would read the value in each element of array_a and add that value to the selected element of array_b. The results would be placed in output array array_c. This algorithm could be used to perform a "digital offset" of signal data. The scalar value does not have to be from a single element array. The scalar value may be from an element within a larger array, selected from the "scalar offset" parameter.

```
scalar_multiply /* pseudo code example */
        input1 = array_a
        scalar = array_b
        scalar_offset = 0
        iteration = 10
        output = array_c;
```

This algorithm would read the value in each element of array_a and multiply that value to the selected element of array_b. The results would be placed in output array array_c. The process is be repeated for each element, limited by the iteration count. This algorithm could be used to perform a "digital amplification" of signal data.

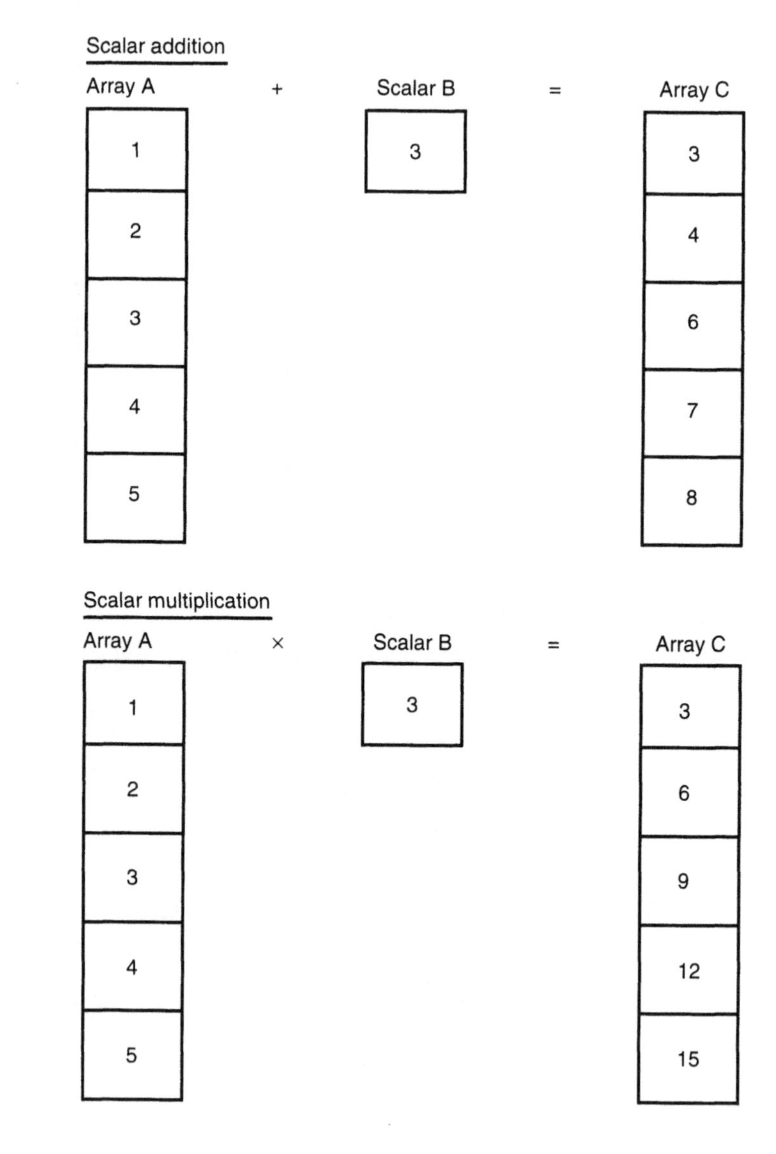

Figure 6.7

Scalar Addition and Multiplication

6.4 Data Type Conversions

Most DSP algorithms are designed for floating-point data. If the input data (from a digitizer or another source) is not in floating-point format, it must be converted to floating-point before processing can take place.

Common data types:

Decimal integer
Binary integer—positive data only
Binary offset—the MSB is the sign bit
Two's complement—the MSB sign bit is inverted
Double precision—64-bit floating point

The DSP format conversion algorithm must specify the input array (of the source data type), the output array (of the destination data type), and the number of iterations. Some method of scale normalization is usually included in the data type conversion sequence. For example, a 16-bit ADC device will generate output codes that range from hex 00 to hex FF. A data type conversion process would be necessary to convert the DUT output data set to a floating-point representation. Once the captured data is in floating-point form, the DSP signal algorithms can be applied to extract test measurement values.

```
/* pseudo code to convert integer data into floating-point data */
int_to_float
   from: dsp array A
         array size = 1024
         input stride = 1
   to: dsp array C
       array size = 1024
       output stride = 1
       iterations = 1024;
```

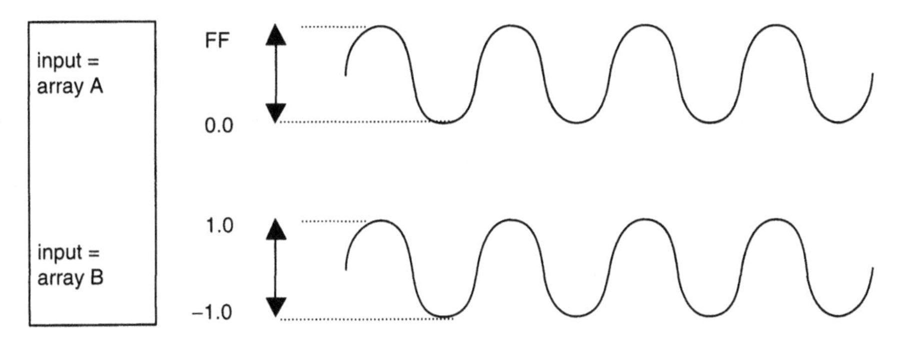

Figure 6.8

Data Type Conversions

6.5 Signal Analysis Algorithms

The true power of the DSP technology for mixed signal test applications lies in the signal analysis algorithms. While a conventional computer is capable of generating the same results, a DSP-based approach is usually faster in terms of execution and development. Anything that can be measured on conventional analog instruments can be measured with the DSP. Test program development can be organized as three distinct objectives. The first objective is to properly condition the device and apply the correct input. The second objective is to capture the device output data, whether analog or digital. The third step is to program the DSP sequence to analyze and evaluate the captured signal characteristics.

6.5.1 Peak and RMS Measurements

Some DSP signal analysis algorithms, such as average and peak measurements, are simple implementations of conventional analog measurements. Additional sophistication is available to significantly extend the ATE system measurement capabilities. The following examples show how the DSP can function as a virtual AC voltage meter by using the min and max functions to analyze arrays containing time domain signal data. The min and max algorithms scan through the input vector, and return the value and location of the specific parameter.

```
max_value /* find the largest value in the array */
input = digitizer_array
      output = max_array
      max_loc = location
      iteration = 128;
      min_value /* find the lowest value in the array */
      input = digitizer_array
      output = min_array
      min_loc = location
      iteration = 128;
      add /* add the min and max to get the peak-to-peak value */
      input1 = max_array
      input2 = min_array
      output = peak_to_peak_array
      iteration = 1;
```

A similar sequence of DSP algorithms could be used to extract the positive and negative peaks, the RMS, and the average (DC offset) of the signal.

6.5.2 The Levels Algorithm

The levels algorithm acts like a software version of a comparator. The output array generated by the levels algorithm contains either a 1 or a 0, corresponding to the result of the comparison for each element.

levels

```
input1 = digitizer_array
input2 = levels_value
output1 = levels_out
iteration = 128;
```

This algorithm compares the value of each element of array input1 with the value contained in variable input2. Each element within the level_out array corresponding to the element within the digitizer_array with a value of 0.5 or less will contain a zero. Each element within the level_out array corresponding to the element within the digitizer_array with a value of more than 0.5 will contain a one.

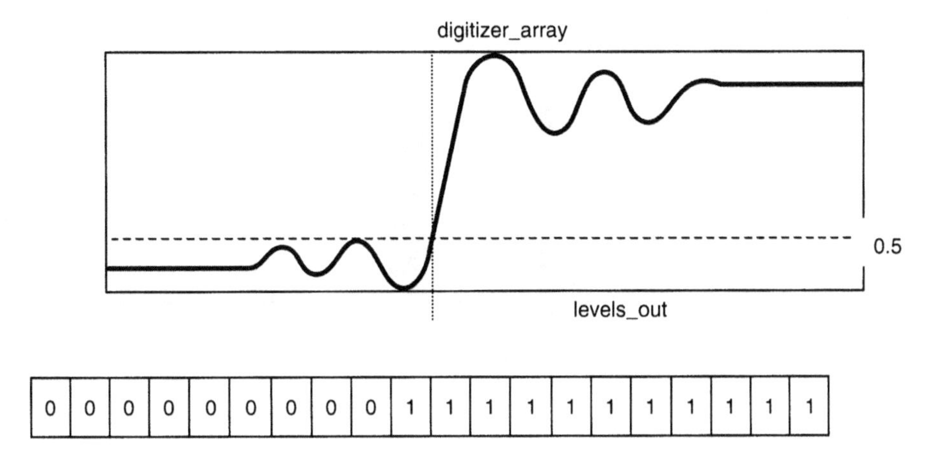

Figure 6.9

Levels Algorithm

6.5.3 The Histogram Algorithm

A histogram is a structure that organizes data according to the number of occurrences of a value. The output of the histogram algorithm can be visualized as a bar chart displaying the number of events corresponding to a measured value.

```
float digitizer_array[128];
float histo_array[128];

      histogram
         input = digitizer_array
         output = histo_array
         max_loc = location
         iteration = 128;
```

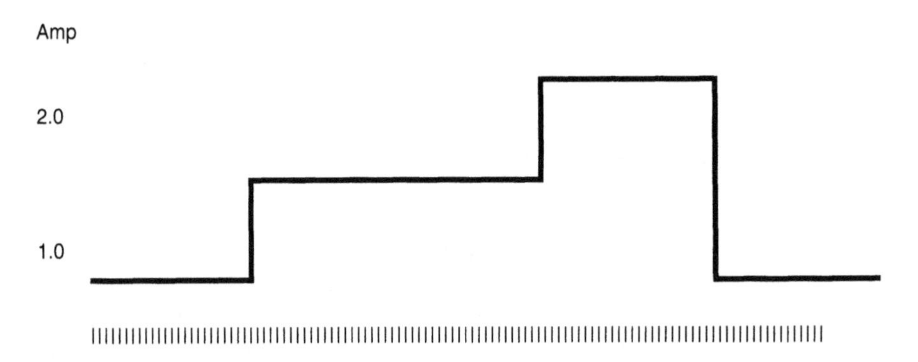

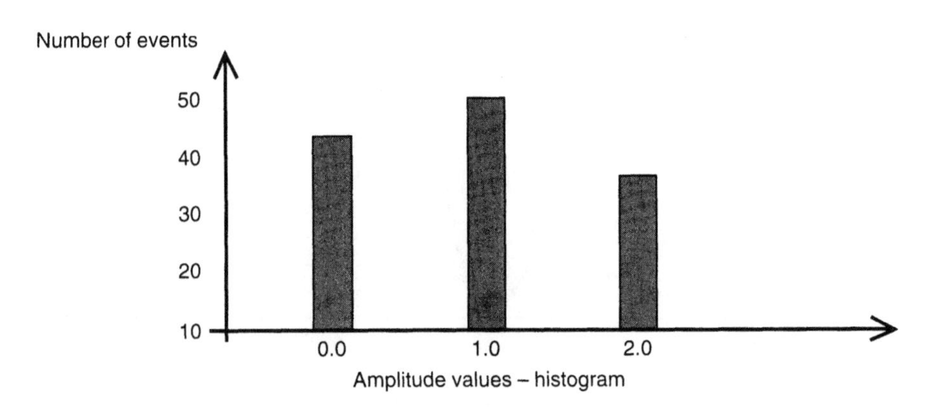

Figure 6.10

Histogram Function

6.6 DSP Measurement Applications

Processing the captured signal data as an array provides tremendous advantages over conventional analog-based instrumentation. The DSP can be programmed to intelligently analyze signal characteristics in both the time domain and the frequency domain.

For example, combining the histogram and level comparison algorithms provides powerful signal analysis capability by allowing adaptive processes. Consider an application that requires DSP techniques to measure rise time. Each device may generate a signal that varies in amplitude and edge location. The DSP algorithm sequence must adapt to variations in the signal levels and timing.

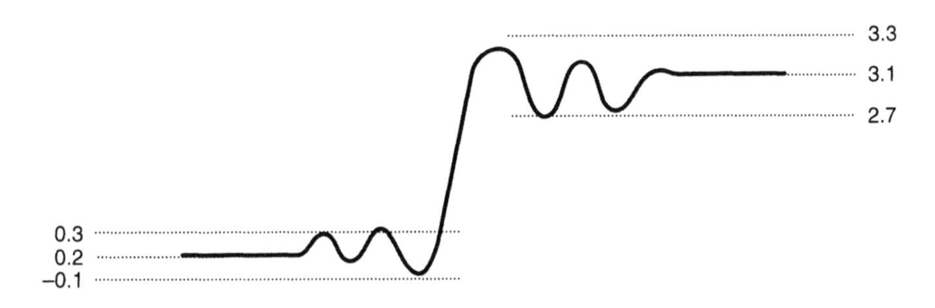

Figure 6.11

Rising Edge Signal Data

6.6.1 Adaptive Rise Time Measurement Example

Execute Histogram: A histogram of the captured signal data set results in a bi-modal distribution, indicating the low and high baselines of the signal.

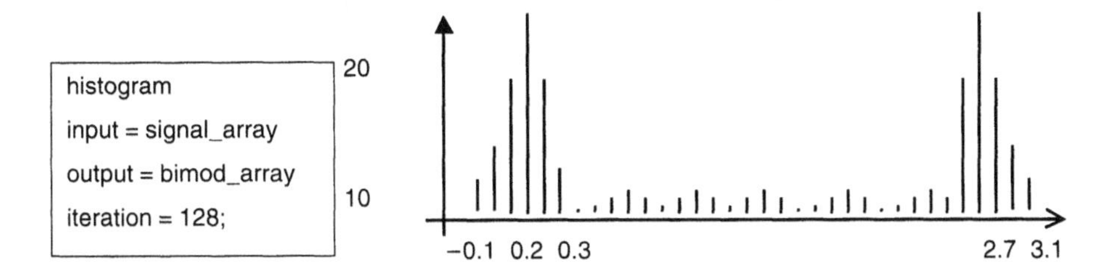

Figure 6.12

Bi-modal Histogram Distribution

Find the first peak: A max_value algorithm locates the element and value of one of the histogram peaks after the peak value is stored, the peak element and the adjacent elements are cleared to prevent a false response when searching for the second peak.

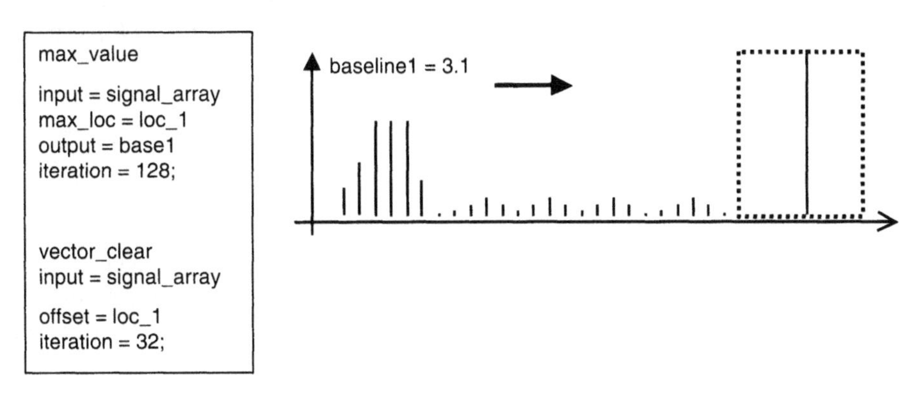

Figure 6.13

Find the First Peak, then Clear Adjacent Values

Find the second peak: A second max_value algorithm locates the element and value of the remaining histogram peak. The peak with the most positive value is assigned to the variable *hi_baseline*, and the other peak is assigned to the variable *lo_baseline*.

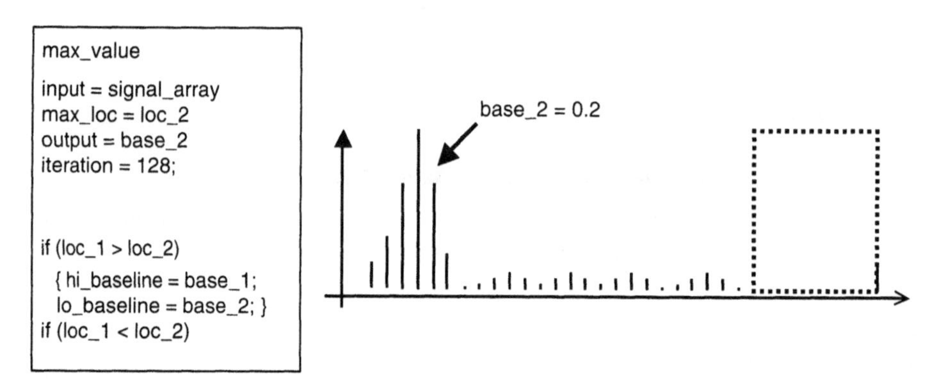

Figure 6.14

Find the Second Peak

Calculate the thresholds: If the first peak value is of greater magnitude than the second peak value, then the first peak is assigned to the hi_baseline, and the second peak is assigned to the lo_baseline. Conversely, if the first peak value is less than the second peak, then the first peak is assigned to the lo_baseline, and the second peak is assigned to the variable hi_baseline. The difference between *hi_baseline* and *lo_baseline* is used to calculate the 10% threshold and 90% threshold, by multiplying the difference by 0.1 and 0.9, respectively.

```
span = hi_baseline - lo_baseline;
thresh_10 = 0.1 * span;
thresh_90 = 0.9 * span;
```

First threshold levels algorithm: A levels algorithm is programmed to a 10% threshold value, (calculated from the histogram data). Executing the levels algorithm generates an array of binary values corresponding to the signal transition at the 10% threshold.

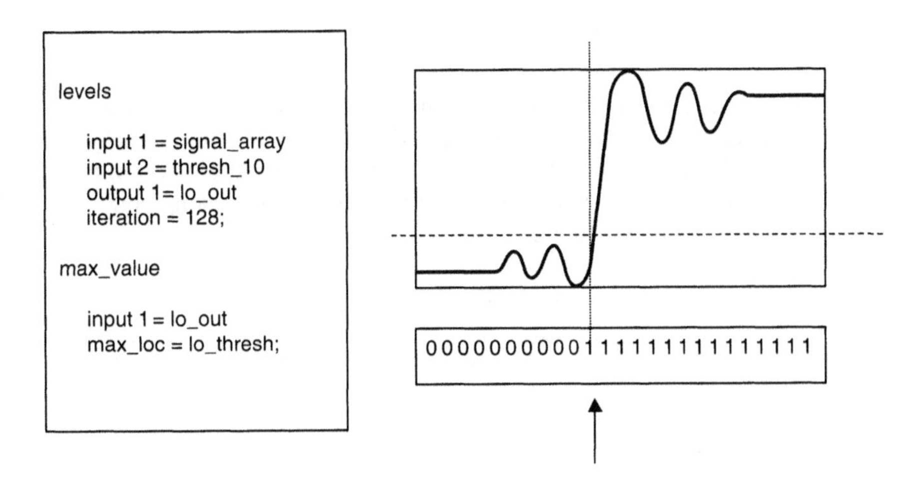

Figure 6.15

10% Threshold Levels Algorithm

Second threshold levels algorithm: A second levels algorithm is programmed to a 90% threshold value. Executing this step returns the value and location of the 90% threshold.

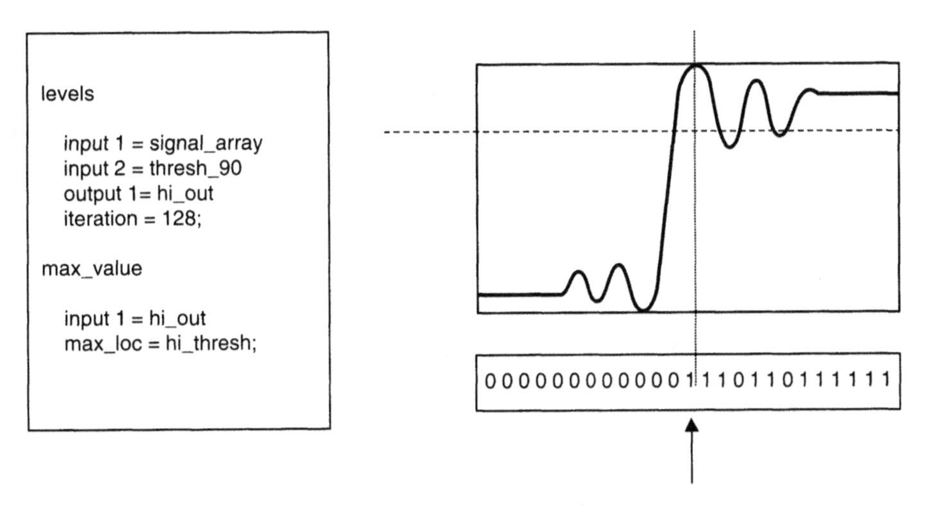

Figure 6.16

90% Threshold Levels Algorithm

Rise time calculation: Subtracting the element locations corresponding to the 10% and 90% levels returns the number of elements between the two thresholds. Multiplying the difference by the sample period produces the rise time value.

```
hi_thresh - lo_thresh = delta;
delta * period = rise_time;
```

6.7 FFT Algorithms

A forward FFT transforms time domain data into frequency domain data. A reverse, or inverse, FFT transforms frequency domain data into time domain data. The input to the FFT algorithm is a floating-point set and requires a power of two sample size. Usually the ATE system DSP algorithm converts the floating-point data into a type complex before executing the FFT transform.

```
complex array fft_complex[16]
float array signal_array[16];
/* converting time domain data in signal_array to frequency data. */
/* the frequency data is type complex (two-dimensional) */
         fft
         input = signal_array
         output = fft_complex
         iteration = 16;
```

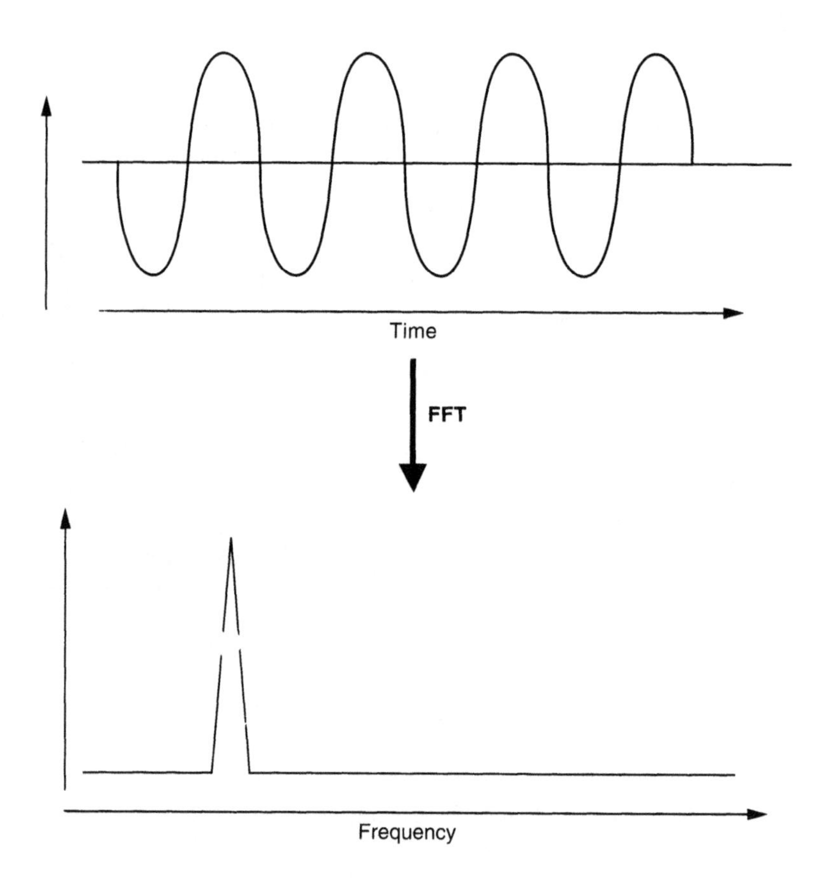

Figure 6.17

The FFT Algorithm

The output of the FFT algorithm is a type complex data set that represents the real and imaginary components of the signal data. Most test applications require converting the FFT complex output into magnitude information.

6.7.1 Processing the FFT Complex Data Set

The output of an FFT can be processed with a magnitude algorithm to generate a floating-point data set that represents frequency domain data. The polar algorithm performs the square root of the sum of the squares, which generates the full data set. The complex vector magnitude algorithm only performs the sum of the squares, r2 + i2. This abbreviated processing is sufficient for relative measurements, and is useful for calculating power ratios (the sum of squares).

The input to the magnitude algorithm is a type complex data set. Because the polar and complex magnitude processes extract a single floating-point value for each complex pair from 0 Hz to Nyquist (fs/2), the output of the magnitude algorithm is one-half the size of the complex input array. The negative frequency is redundant, so only the positive frequency is processed.

```
float fft_complex [2][16]; /* two dimensional array for type complex */
float mags_array[8];
/* convert complex data into a floating point magnitude data. */
cvmags
          input = fft_complex
          output = mags_array
```

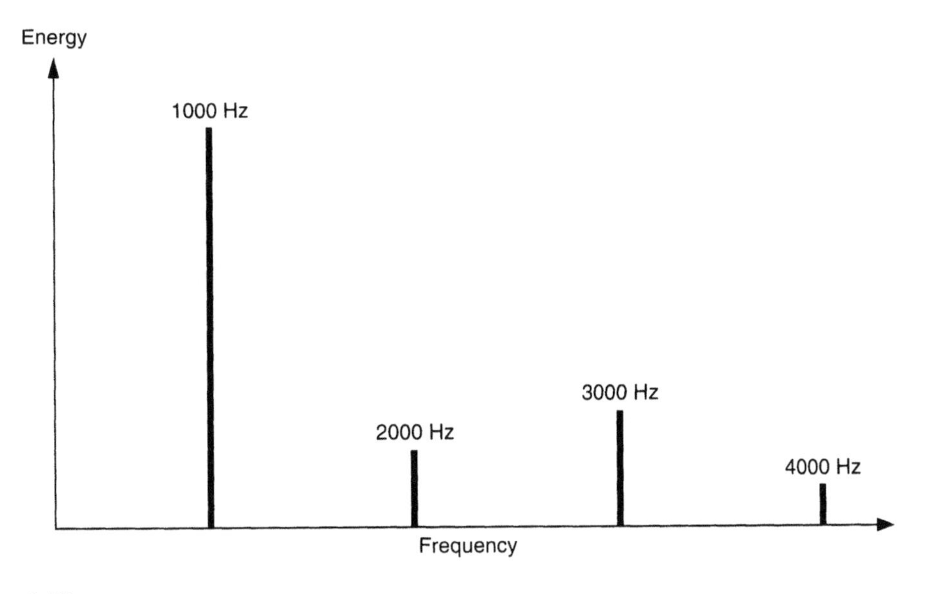

Figure 6.18

Frequency Domain Data

6.7.2 The Inverse FFT

Converting from the frequency domain into the time domain has several useful applications, including brick wall filters and mathematical oversampling. The DSP process to execute an inverse FFT is similar in structure to the forward FFT.

```
complex array fft_complex[16]
float array signal_array [16];
/* converting frequency domain data fft_complex into time domain data. */
/* the frequency data is type complex (two-dimensional) */
        ifft
        input = fft_complex
        output=signal_array
        iteration = 16;
```

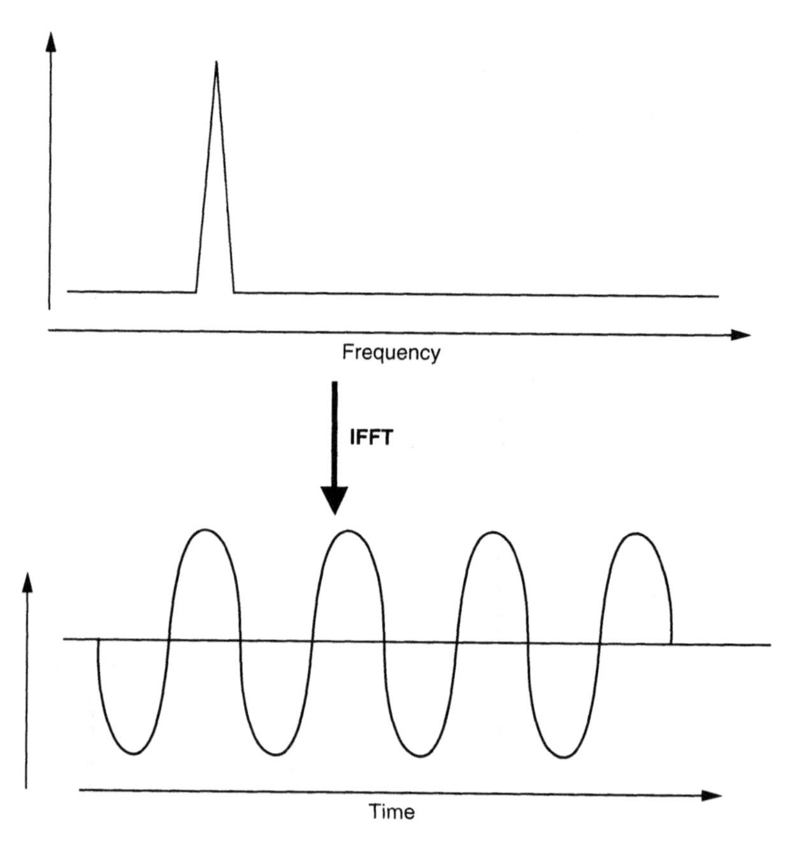

Figure 6.19

The Inverse FFT Algorithm

Multi-tones in the Frequency Domain

The inverse FFT has some interesting applications, including the generation of multi-tone waveform arrays. Rather than building the multi-tone signal as a sum of time-domain data sets, it's sometimes easier to create the multi-tone data in the frequency domain, and then perform an inverse FFT to generate the time domain data set.

Recall that a multi-tone signal in the frequency domain will produce distinct line spectra for each tone in the signal, and that each signal component is of equal amplitude. What we're going to do is start with the frequency domain description, and then work backward to produce the multi-tone signal array. Let's start with a two-dimensional floating-point array, called f_domain. One part of the array is used for the magnitude information, while the other part is used for the phase information.

Magnitude																															
Phase																															

Figure 6.20

Type Complex Signal Data Array

Because you are working in the frequency domain, you can apply all that cool stuff from Chapter 4 about the frequency bin. In the frequency domain, the frequency bin (fbin) number correlates directly to the location of the signal frequency. Revisiting the examples from Chapter 4, let's say you've got an fbase of 120 Hz, and the target multi-tone is composed of 960, 2040, and 3120. The corresponding fbin values are determined by

$$\text{fbin} = \frac{\text{fi}}{\text{fbase}}$$

Sounds familiar, right? All you need to do in the frequency domain is to set the values of the target fbin elements to 1.0 and clear every other element. You don't want any signal energy at frequencies other than the target tones, so all the other elements are set to zero. If you plotted this out, guess what? You get a frequency domain plot of the multi-tone waveform.

Mag.	0	0	0	0	0	0	0	1	0	0	0	0	0	0	0	0	1	0	0	0	0	0	0	0	0	1	0	0	0	0	0
Phase																															

Figure 6.21

Entering Magnitude and Phase Information

For the phase array, you can enter the phase value (in radians) for each tone in the corresponding fbin location. For this example, let's pick some random values between 0.0 and 6.28 (2 × PI). The phase information in the remaining elements can be set to zero.

We're all set! Now that the data has been initialized, the next step is to convert the polar form into type complex. The "polar_complex" algorithm of our dream-machine DSP should do the trick.

```
polar_complex
      input: f_domain[0] /* magnitude */
      input2: f_domain[1] /* phase */
      output1: complex_array
      iteration: 16
```

Once the frequency domain data set is in type complex, it's ready for an inverse FFT. Here it is:

```
ifft
input = fft_complex_array /* frequency domain */
output = signal_array     /* time domain */
iteration = 16;
```

This is so cool, I love it when a plan comes together. Guess what you get in the algorithm output array, signal_array? You betcha! Here's your multi-tone.

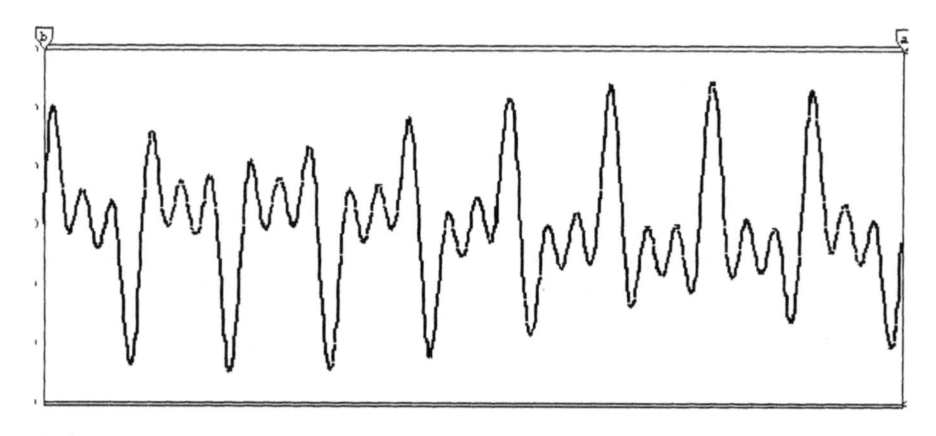

Figure 6.22

Creating a Multi-tone with an Inverse FFT

Other Cool Stuff

The DSP algorithm library of the ATE system is like a big toolbox. Some of the algorithms may be the equivalent of a garage-sale tool that looks like a left-handed grommet extrusion reversal wrench, but most of the goodies are pretty useful. Here's a quick tour of what you might find in your dream machine DSP.

Reverse (input1, output1, size)

Action: Reverses the contents of an array from left to right. This is useful for some trick stuff with settling time measurements, as you'll see in the chapter about DAC testing.

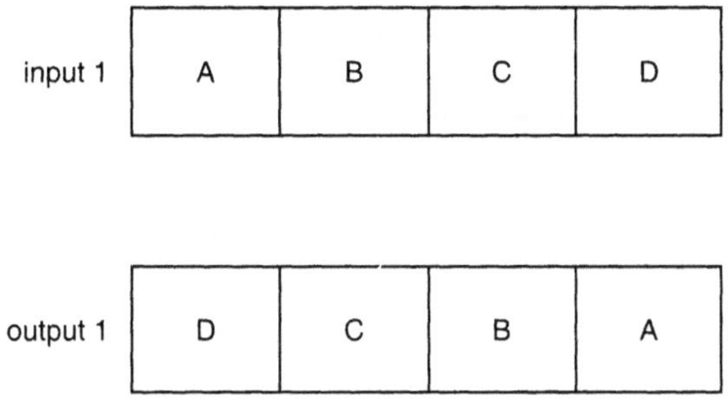

input 1	A	B	C	D
output 1	D	C	B	A

Figure 6.23

Reverse Order Algorithm

Copy and Subtract (input1, output1, size)

Action: First, the algorithm uses an offset to copy the contents of input1 to a temporary array. Second, the contents of the temporary array are subtracted from the input1 array to create a first-order differentiation. This is useful for identifying change-of-slope transitions.

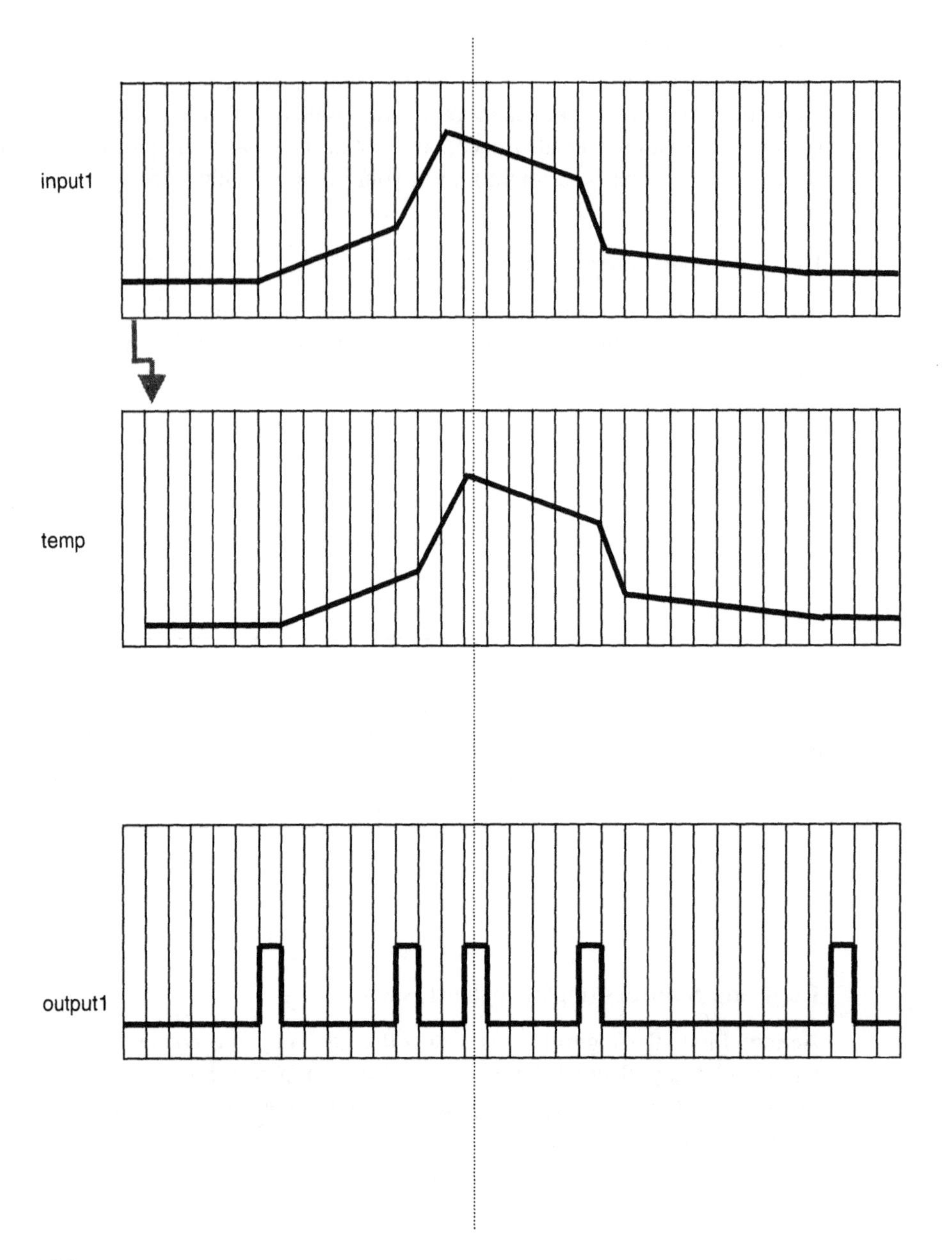

Figure 6.24

Copy, Shift, and Subtract = Differentiation

Shift and Rotate (input1, offset, size, direction)

Action: The algorithm shifts the contents of the input1 array by the number of elements specified in the "offset" parameter. The "direction" flag determines if the algorithm performs a left-shift or a right-shift of the array elements. For applications such as template testing, it is sometimes useful to shift the effective phase of the captured signal data.

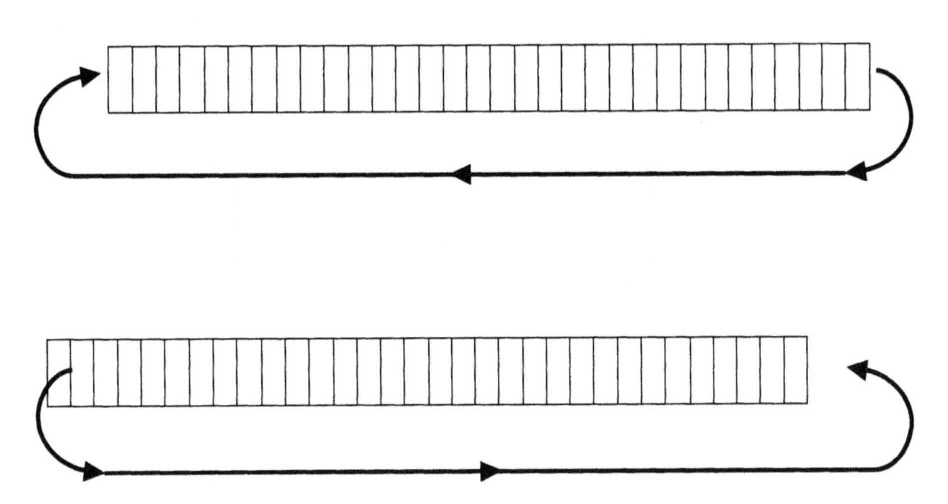

Figure 6.25

Shift and Rotate = Phase Adjustment

6.8 Harmonic Distortion Test Sequence

Testing for harmonic distortion requires a comparison (in dB or percentage) between the energy for the fundamental frequency, and the energy in the harmonics of the fundamental. The DSP process must determine the energy value of the fundamental by reading the value at the frequency location in the frequency domain. Signal energy corresponding to integer multiples of the original signal frequency are summed together into the harmonic distortion level. The noise (non-harmonic energy) and the DC component (0Hz) are ignored.

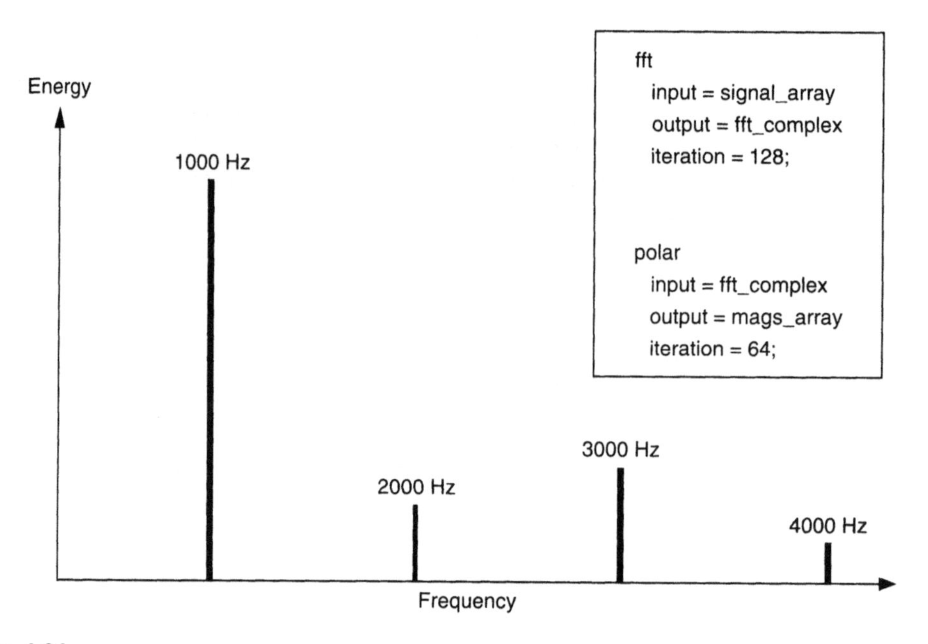

Figure 6.26

Harmonic Distortion in the Frequency Domain

6.8.1 Extracting the Fundamental and Harmonics

Because the frequency domain data set is simply an array of energy values arranged in increments of fbase, we can use the following equation to locate the array element containing the fundamental signal energy:

$$\text{fbase} = \frac{\text{fs}}{\text{samples}} \qquad \text{array\_location} = \frac{\text{fi}}{\text{fbase}}$$

First we would calculate the base frequency, fbase. Once the fbase value has been identified, the array location of any frequency component can be calculated and extracted. The harmonics energy locations in the frequency domain array will be integer multiples of the fundamental frequency location. The harmonic energy values are algebraically summed to mathematically model the self-canceling effect of even-order and odd-order harmonics.

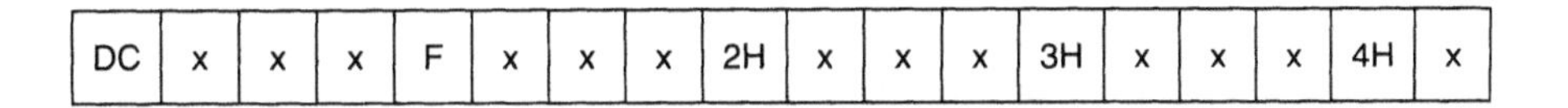

$$\text{Total\_Harmonic\_Energy} = \sqrt{2nd^2 + 3rd^2 + 4th^2 \ldots}$$

$$\text{Total\_Harmonic\_ Distortion} = \frac{\text{Total\_ Harmonic\_Energy}}{\text{Fund}} \times 100$$

Figure 6.27

Extracting the Fundamental and Distortion Components

6.8.2 DSP Pseudo Code Example

```
/* DSP Sequence to measure harmonic distortion
/* fs was 128 kHz, samples = 1024, fbase = 125 Hz */
float signal_array[1024]; complex fft_array[1024];
float mags_array[512]; THD;THDSQ; THD_ratio;
float THD_percent[1]; fundamental[1]; float harmonic_energy[4];

fft /* do the FFT */
        input = signal_array /* sample rate was 128kHz */
        output=fft_array
        iteration = 1024;
```

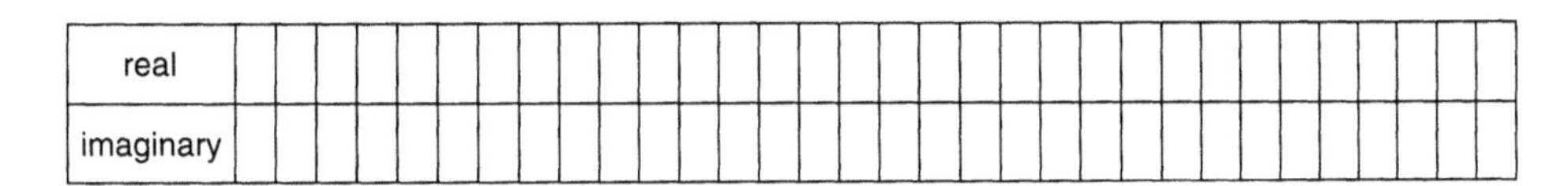

Figure 6.28

Type Complex FFT Results

```
polar /* polar conversion of raw FFT data */
      input = fft_complex
      output= mags_array
      iteration = 512;
```

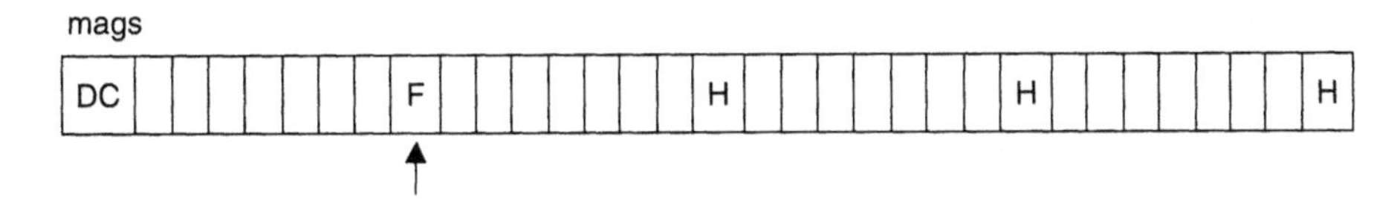

Figure 6.29

Locating the Fundamental

```
/* get the fundamental frequency energy */
copy
   input1: mags_array
   offset=8          /* frequency_location = fi/fbase = 1000Hz/125Hz*
   output=fundamental
   iteration=1

/* bump through every 8th bin to get harmonic energy sum */
copy
   input1=mags_array
   stride=8
   offset=16 /* start at second harmonic */
   output=harmonic_energy;
   iteration=3;
```

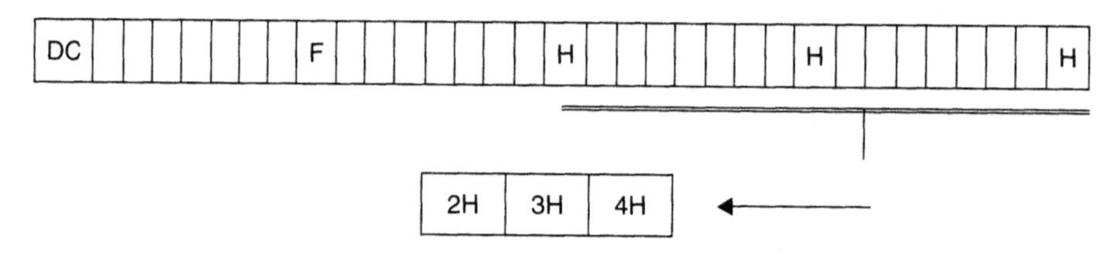

Figure 6.30

Processing the Harmonic Data

```
/* square each harmonic */

vector_square /* square harmonics 'in place', and re-use the array */
     input1=harmonic_energy
     iteration = 3
     output1=harmonic_energy
```

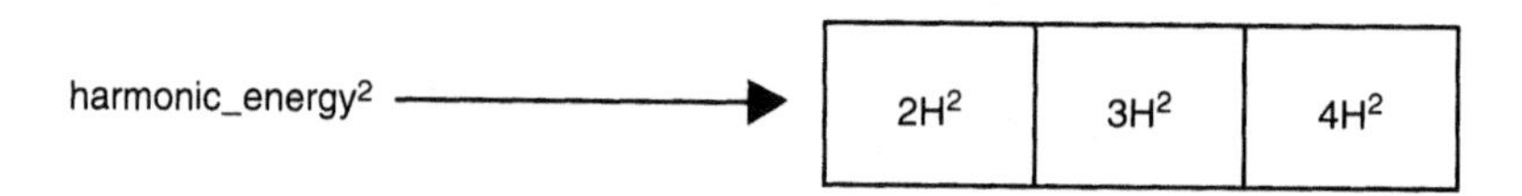

Figure 6.31

Calculating the Root Mean Square

```
/* get sum of the squares */
      vector_sum
        input1 = harmonic_energy
        iteration = 3
        output = THD;
```

Figure 6.32

Summing the Harmonic Energy

```
/* get square root of harmonic energy */
        square_root
          input1 = THD
          iteration = 1
          output = THDSQ;

/* divide the distortion energy by the fundamental energy */
        scalar_divide
          input1 = THDSQ
          input2 = fundamental
          output = THD_ratio /* get the ratio of fund to distortion */
          iteration = 1;

/* calculate the percentage */
        scalar_multiply
          input1 = THD
          input2 = 100
          iteration = 1
          output = THD_percentage; /* all done! */
```

6.8.3 Noise Measurement

Okay, so much for harmonic distortion. What about noise testing? Well, instead of summing the harmonic energy, noise testing sums all of the signal energy, excluding the DC component, the fundamental, and the harmonics. The square root sum of squares of the remaining signal energy represents the noise figure.

```
/* DSP Sequence to signal to noise */
/* fs was 128 kHz, samples = 1024, fbase = 125 Hz */
float signal_array[1024]; complex fft_array[1024];
float mags_array[512]; float noise_energy; noise_total, noise_sq; SNR_log; SNR;
fft /* do the FFT */
            input = signal_array /* sample rate was 128kHz */
            output = ft_array
            iteration = 1024;
```

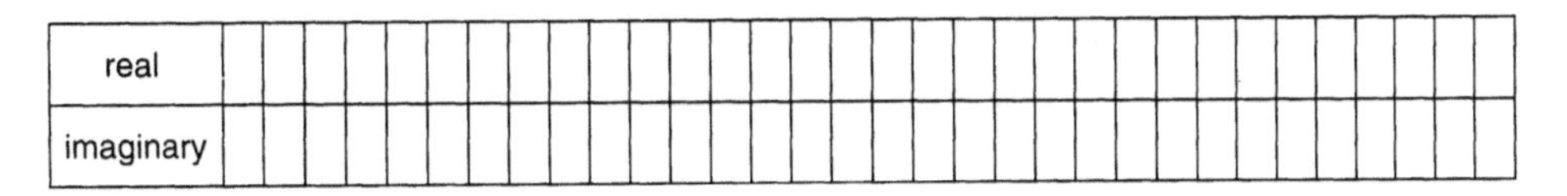

Figure 6.33

Noise Measurement Data

```
polar /* polar conversion of raw FFT data */
      input = fft_complex
      output = mags_array
      iteration = 512;
```

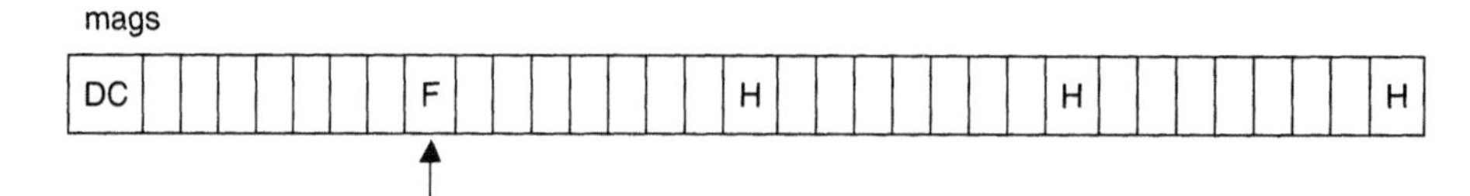

Figure 6.34

Find Fundamental, then Remove DC and Distortion

```
/* get the fundamental frequency energy */
     copy
        input1:mags_array
        offset=8          /* frequency location = fi / fbase = 1000Hz / 125Hz */
        output=fundamental
        iteration=1

/* write a zero into every 8th bin to remove DC, fundamental, and harmonic
  energy */
     clear
         input1 = mags_array
         stride = 8
         offset = 0 /* start at DC */
         output = noise_energy;
         iteration = 5;
```

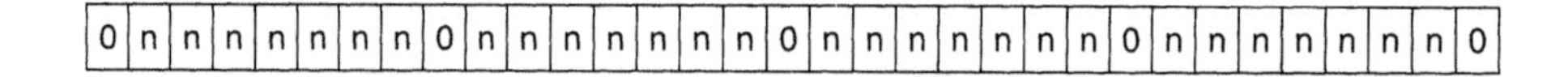

Figure 6.35

Calculate Square Root Sum of Squares for Noise Components

```
/* square each noise component */
vector_square /* square harmonics 'in place', and re-use the array */
             input1 = noise_energy
             iteration = 3
             output1 = noise_energy

/* get sum of the squares */
     vector_sum
          input1 = noise_energy       noise_total
          iteration = 3
          output = noise_total;
```

```
/* get square root of noise energy */
        square_root
           input1 = noise_total
           iteration = 1
           output = noise_sq;

/* divide the noise energy by the fundamental energy */
        scalar_divide
           input1 = noise_sq
           input2 = fundamental
           output = SNR_ratio /* get the ratio of signal to noise */
           iteration = 1;

/* calculate the log ratio */
        log
           input1 = SNR_ratio
           output = SNR_log;

/* scale by - 20 */
        scalar_multiply
           input1 = SNR_log
           scalar = -20
           output = SNR;      /* all done! */
```

Chapter Review Questions

Sample Array Set

Array 'A'	Array 'B'	Array 'C'
1	2	
2	3	
3	2	
4	3	

Figure 6.36

Knowledge Check—Example DSP Arrays

1. After executing this algorithm, what will be the contents of array C?

```
vector_add
     input_1 = array_A
     input_2 = array_B
     output = array_C
     iteration = 4;
```

2. After executing this algorithm, what will be the contents of array C?

```
scalar_add
     input_1 = array_A
     scalar = array_B
     offset = 0
     output = array_C
     iteration = 4;
```

3. Match these four time domain signal characteristics with the associated DSP algorithms that would be used to make the measurement.

Time Domain Measurement	**DSP Algorithm**
DC Offset	FFT
Negative Peak	RMS
Signal Energy	HISTOGRAM
Positive Peak	MAX
	MIN
	REVERSE
	AVERAGE

CHAPTER 7

TESTING DIGITAL-TO-ANALOG CONVERTERS

Trifles make perfection, but perfection is no trifle.
—Michelangelo

7.1 Introduction

The digital-to-analog converter (DAC) is perhaps the most ubiquitous type of mixed signal device. Applications for DACs span a wide range, from consumer audio to precision instrumentation, and everything in between. Whether or not you spend much of your career testing DACs, it's still a good idea to know how. There is a terminology and methodology associated with DAC tests that applies to many other areas of mixed signal test. Of course, not all DAC types are tested for all parameters. Instead, a set of basics tests common to DACs in general are supplemented with specific tests targeted toward the end-use application.

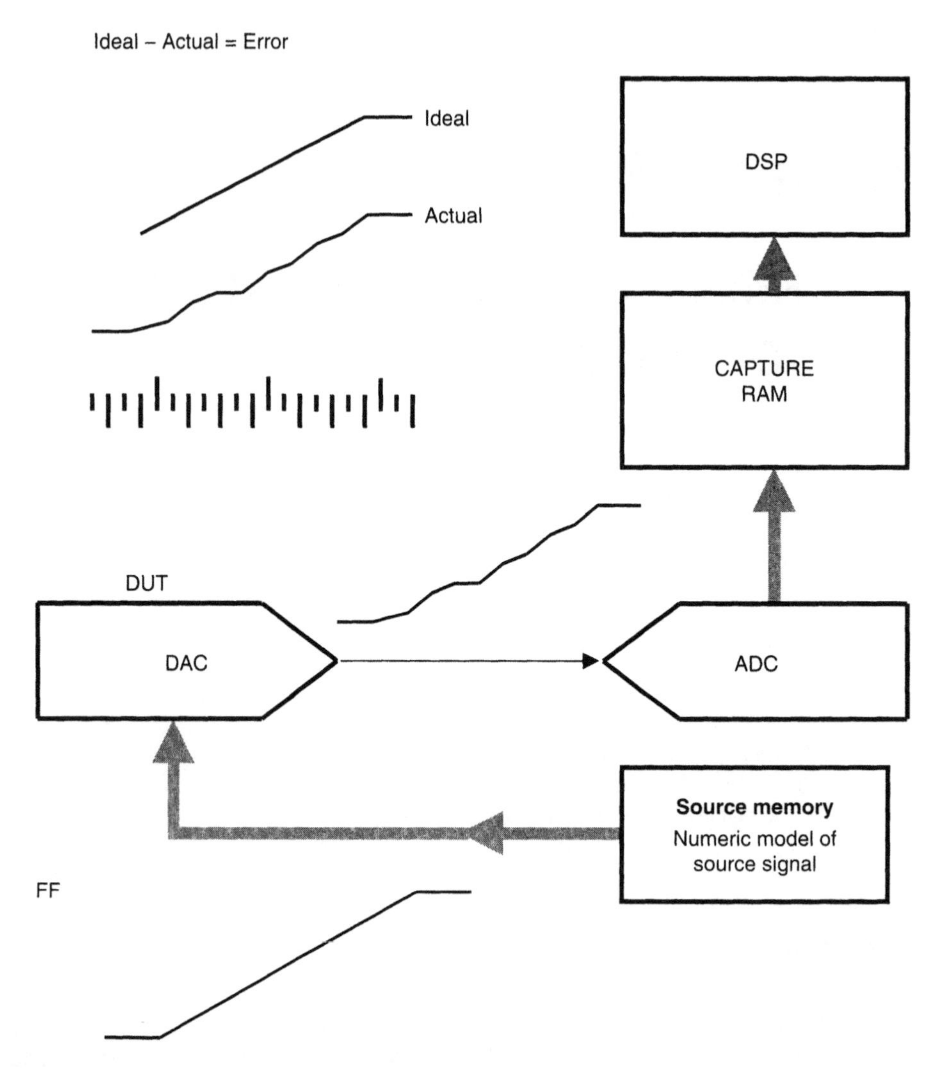

Figure 7.1

DAC Test Overview

A DAC produces a set of analog values that corresponds to the digital input codes. Many DAC designs are actually current output devices, instead of voltage output devices. More often than not, we'll convert the output current to an equivalent voltage level, so the examples in this chapter are based on a voltage output converter.

Testing the DC performance of a DAC consists largely of verifying a consistent and linear response. A typical test setup for testing the DC performance of a DAC device uses the test system signal source to generate a ramp. This ramp signal is routed to the digital pin electronics, which applies voltage level and timing formats. The DAC is driven with sequential digital vectors, causing the DAC analog output to generate a voltage ramp.

Efficient analysis of the device response requires that the output voltage be digitized and then processed by the test system's Digital Signal Processor (DSP). The DSP is used to subtract the digitized DAC output from a calculated ideal. The difference between the ideal and actual signal data is analyzed to evaluate DC performance.

In order to test the accuracy of each DAC output voltage, the resolution of the test system digitizer must be much greater than the resolution of the DAC under test. If the DAC under test is 12 bits, the digitizer must have an additional 4 bits to achieve a linearity measurement accuracy of 1/16 of the DAC LSB. Measuring to 1/16th of the DAC LSB results does not provide a 5% resolution of each step value. For linearity testing, the rule of thumb for testing DACs is that the test system digitizer must have a resolution of at least 2 bits greater than the resolution of the DAC, or 1/4 of an LSB. An additional 3 bits of measurement precision is a practical minimum.

Example

The DUT is a 12-bit DAC with a 10-volt range. An ideal approximation of the step size is

$$\frac{10.0 \text{ v}}{2^n} = \frac{10.0 \text{ v}}{4096} = 2.44 \text{ mV}$$

Measuring the DUT output with a 16-bit measurement system (an additional 4 bits) allows a measurement step size of 152.5 μV. This is equal to 6.25% of the DAC step size.

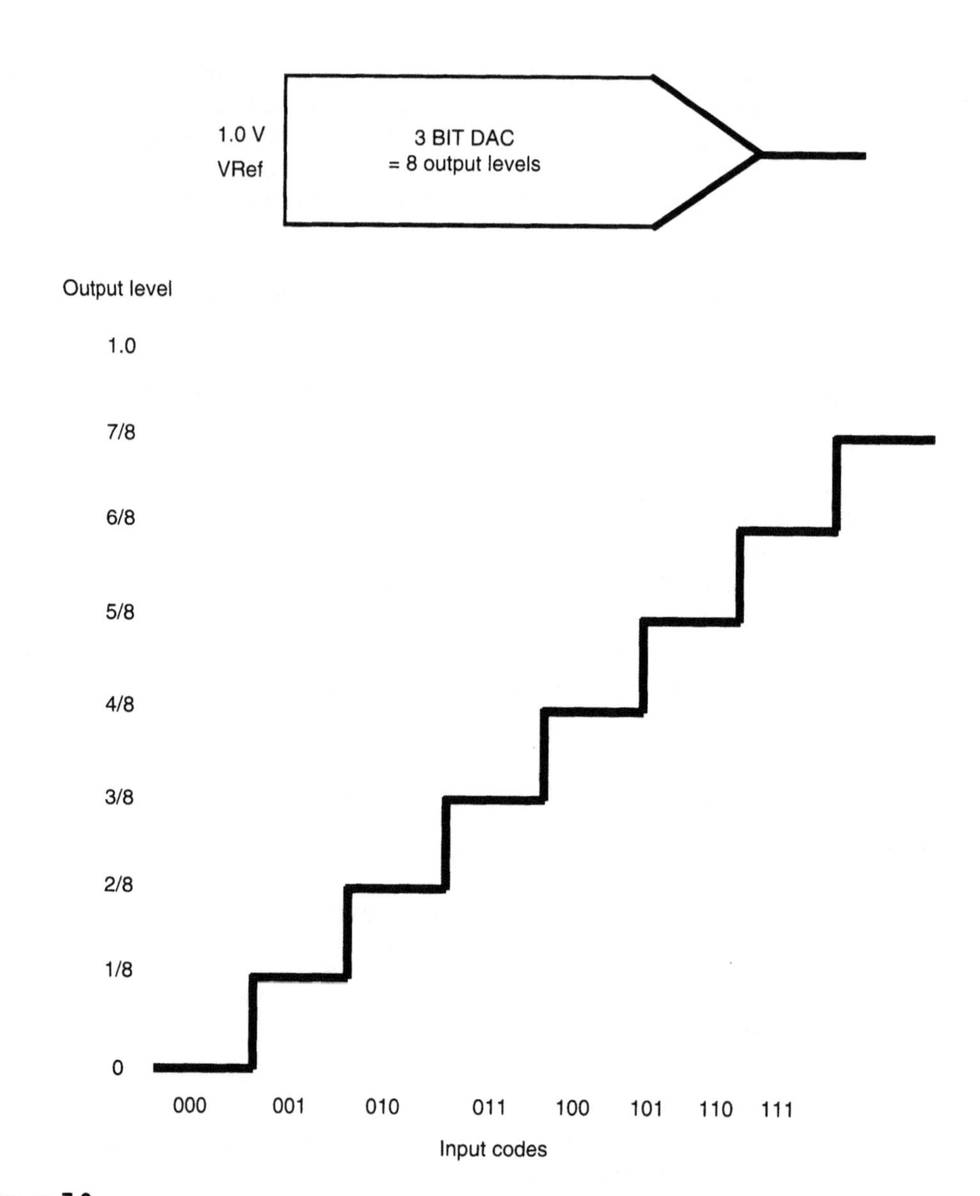

Figure 7.2

DAC Transfer Curve

7.2 DAC Overview

In general, a DAC is specified according to the number of bits (n), and the reference level, also known as the full-scale range (FSR). For a unipolar DAC, the reference level is the same as the reference voltage (VRef). For a bipolar DAC, the reference level is equal to the difference between the positive and negative VRef levels (+VRef and –VRef). Contrary to what is implied by the naming convention, the output of a unipolar digital-to-analog converter does not range from zero volts to the reference level. Actually, the output of a perfect unipolar DAC is one LSB step *less* than the reference voltage. (Don't shoot me! I'm just the piano man!) To keep from passing on a legacy of total confusion, this chapter will use the term reference level. But, keep in mind that this value is sometimes called the FSR.

Consider a 3-bit DAC, with a reference voltage of 1.0 volt. Ideally, we could say that the output of the DAC ranges from 0.0 volts to 1.0 volt, and that each increment of the input code will cause an increase of 1/8 of a volt. The first code causes an output of 0.0 volts, and each increment in the input codes causes the output to increase by 0.125 volts (1/8 of a volt). The full-scale output of the DAC will be 0.875 (7/8 of a volt). That is, at the maximum input code value of 111, the output will be 1 LSB step size less than the reference voltage.

When measuring the step size, keep in mind that the lowest input code does not produce the first step, only the first level. The total number of input codes is equal to 2^n, but the total number of output level steps is equal to $2^n - 1$.

Number of unique input code values = 2^n.
Number of unique output levels = 2^n
Number of output steps = $2^n - 1$

$$\text{Maximum output level (nominal)} = \text{voltage reference} \times \left(\frac{2^n - 1}{2^n} \right)$$

The examples in this chapter use the following nomenclature:

V[n]

where V is the analog output voltage, and n is the digital input value.

7.3 DC Test Overview

There are two general categories of DC tests for DAC devices. The first category evaluates the device minimum and maximum output levels, referenced to an absolute specification. Offset measures the minimum output level, while gain measures the overall output span from the minimum to the maximum.

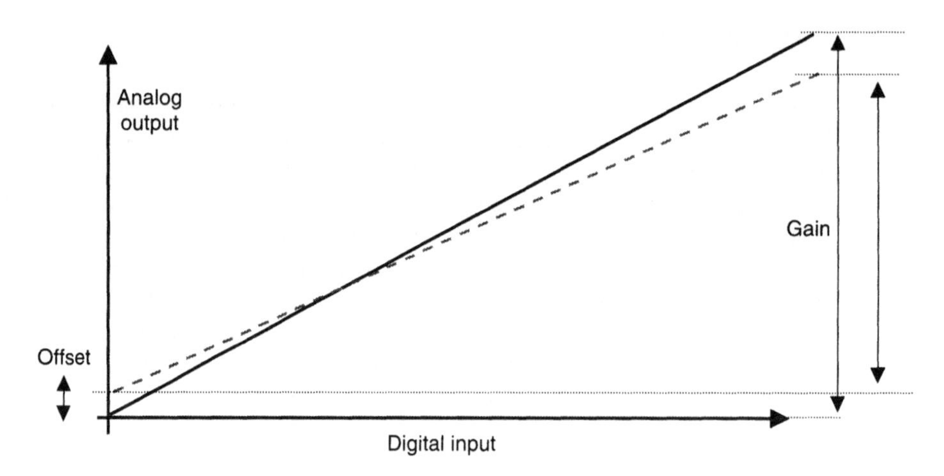

Figure 7.3

Offset and Gain

The second category of DC test evaluates the device linearity, according to a relative step size value that is calculated per device. Because of process variations, the overall analog output span may exhibit variations from one device to another. Two otherwise identical devices may be perfectly linear, but with different endpoints and step size values.

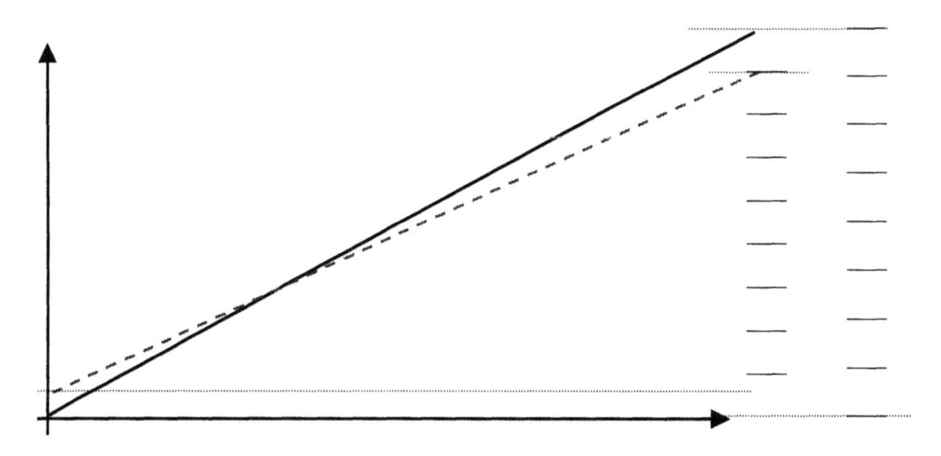

Figure 7.4

DAC Linearity

7.3.1 Offset Measurement

DAC offset is the difference between the ideal and actual analog output for a "zero code" digital input. Some devices have a correction circuit to adjust offset voltage, but offset may be tested as a worst-case measurement without the correction circuit active. Testing for offset consists of measuring the analog output generated by the "zero code" digital input and comparing the value against acceptable limits. Offset may be specified as a voltage, a fraction of an ideal LSB step, or a percentage of the ideal reference level or FSR.

Example

A 3-bit 1-volt DAC is tested for offset error by applying an "all-zeroes" input code to the device digital inputs. The output of the DAC is measured at 10 mV.

Voltage Reference Method

1. Apply "zero code" digital input and measure the output.
2. Compare the measured value against specified limits.

Using the voltage reference method, the measured value of 10 mV is compared against limits of +/–50 mV (as an example, actual applications will vary!). Comparing against limits, –50 mV > 10 mV < 50 mV.

Ideal LSB Reference Method

1. Apply "zero code" digital input and measure the output.
2. Calculate the ideal LSB voltage step as REFERENCE LEVEL/2^n.
3. Calculate the fractional LSB = Voltage[0]/Ideal LSB.
4. Compare the calculated value against specified limits.

According to the ideal LSB reference method, the measured value is compared against limits based on a fraction of an ideal LSB step. The device REFERENCE LEVEL divided by the number of codes (2^n) is equal to 1.0 v/8 = 0.125 volts. The offset error is equal to:

error/LSB = 10 mV/125 mV = 0.08 of an LSB.

Full-Scale Percentage Method

1. Apply "zero code" digital input and measure the output.
2. Calculate the percentage as = (Voltage[0]/Reference Level) × 100.
3. Compare the calculated value against specified limits.

According to the full-scale percentage method, the measured value is compared against limits based on a percentage of the reference level (FSR).

error/Reference Level = 10 mV/1.0 × 100 = 1%

Offset Test

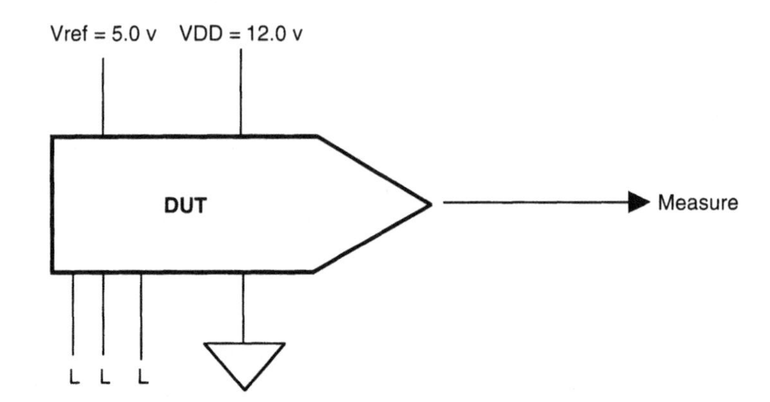

Figure 7.5

Offset Test Circuit

1. Apply power to the device power pins.
2. Apply the voltage reference level to the VREF pin.
3. Set the VIL and VIH level for the digital input pins.
4. Apply the digital input code corresponding to zero voltage output.
5. Measure the analog output level and evaluate the results.

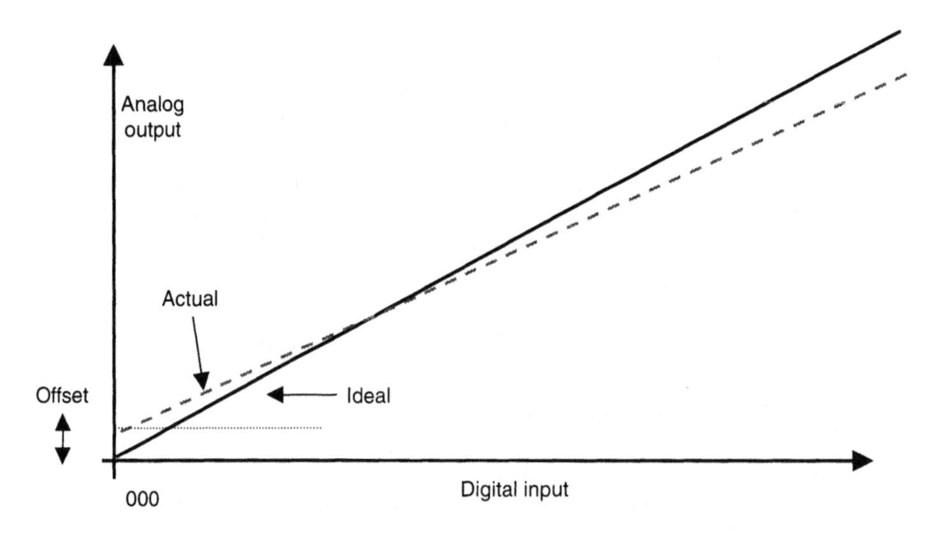

Figure 7.6

Offset Test Plot

7.3.2 Gain Measurement

Gain is the difference between the ideal and actual *span* of analog output values corresponding to the full range of digital input codes. Because the total number of analog steps is $2^n - 1$, the ideal span is equal to the reference level (FSR), minus 1 ideal LSB step. The ideal LSB is equal to reference level/2^n, so the span can be calculated as

$$\text{Reference\_Level} \times \left(\frac{2^n - 1}{2^n}\right)$$

To measure the output span, the test process must measure both the minimum output level (at "all zeroes"), and the maximum output level (at "all ones"). The difference between the minimum and maximum output levels is specified as the device gain. DAC gain can be either positive (greater than the ideal), or negative (less than the ideal).

Example

1. Ideal input span is calculated as

$$\text{Reference\_Level} \times \left(\frac{2^n - 1}{2^n}\right) = 1.0 \text{ volts} \times \left(\frac{7}{8}\right) = 0.875 \text{ volts}$$

2. The measured span is calculated from the measured output levels.
   Measured last code output = 0.870 volts
   Measured zero code output = 0.01 volts
   Measured span = 0.870 – 0.01 = 0.860 volts

3. The gain error is calculated as a percentage of the ideal span.
   Measured span = 0.860 volts
   Ideal span = 0.875 volts

$$\text{Gain} = \frac{\text{Measured\_Span}}{\text{Ideal\_Span}} = \frac{0.860}{0.875} = 0.982$$

4. Calculate the percentage as: Gain – 1.0 × 100

$$0.982 - 1.0 = -0.171$$

$$-0.171 \times 100 = -1.71\%$$

Gain Error Equation

$$\text{GainError} = \left[\left(\frac{V[2^n] - V[0]}{\text{Reference\_Level} \times \left[\frac{2^n - 1}{2^n}\right]}\right) - 1\right] \times 100$$

Gain Test

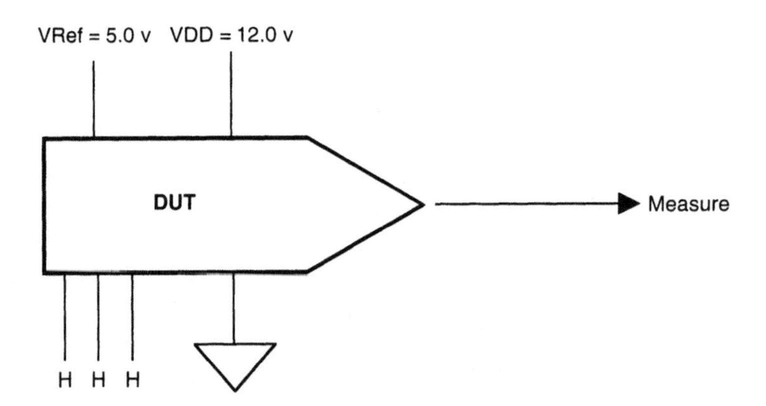

Figure 7.7

Gain Test Circuit

1. Apply power to the device power pins.
2. Apply the voltage reference level to the VREF pin.
3. Set the VIL and VIH level for the digital input pins.
4. Apply the digital input code corresponding to the maximum voltage output.
5. Measure the analog output level.
6. Subtract the measured offset value from the measured maximum output level.
7. Calculate the gain error and evaluate against test limits.

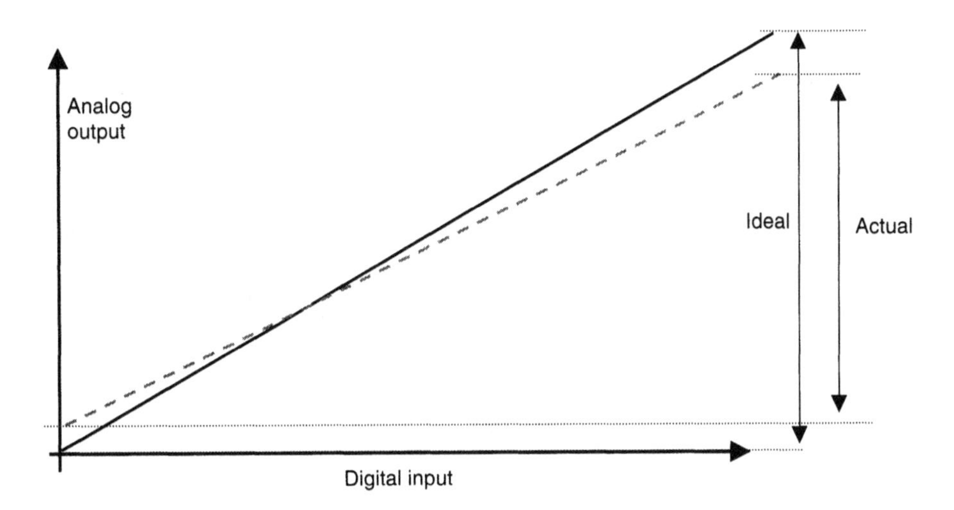

Figure 7.8

Gain Test Plot

7.4 Linearity Test Overview

Ideally, each step of the DAC digital input value would increment the DAC analog output by exactly one step. In an actual device, the analog step size varies. The range of the DAC describes the range of values from the minimum digital code (offset) to the maximum digital code (full scale). Linearity tests evaluate the transfer function (input codes to voltage steps) based on the measured endpoints. The device LSB step is calculated by dividing the measured span of the DAC by the number of possible input codes, as follows:

(Voltage[*n*] indicates the DUT analog level with an input code of *n*.)

$$\text{DeviceLSB} = \left(\frac{\text{Voltage}[2^n - 1] - \text{Voltage}[0]}{2^n - 1} \right)$$

Differential Nonlinearity (DNL) is the difference between the actual step size and the calculated step size. Integral Nonlinearity (INL) is the worst-case variation in any of the analog output values with respect to an ideal straight line drawn through the endpoints. INL is also sometimes defined in comparison to a "best fit" straight line.

If only DNL is tested, then each step could be "in spec," but the accumulated error could be excessive as compared to a straight-line transfer curve. Consider an example device with a DNL specification is +/–25%. If the first ten steps each have a DNL error of 10%, then even though each step is within specification, the accumulated error would be excessive. If only INL is tested, then the device response according to a straight-line transfer curve could be acceptable, but each code step could exhibit excessive variation.

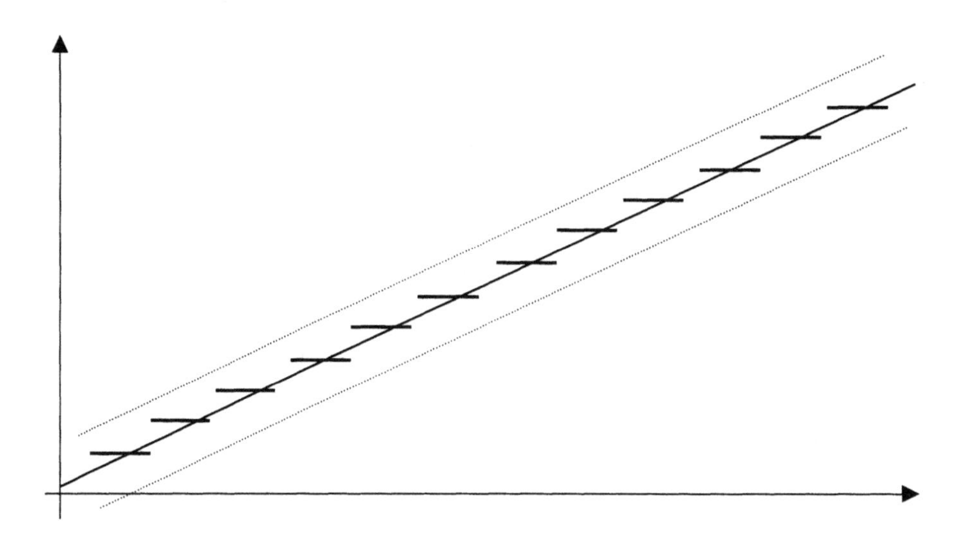

Figure 7.9

DAC Linearity—DNL and INL

Linearity Tests

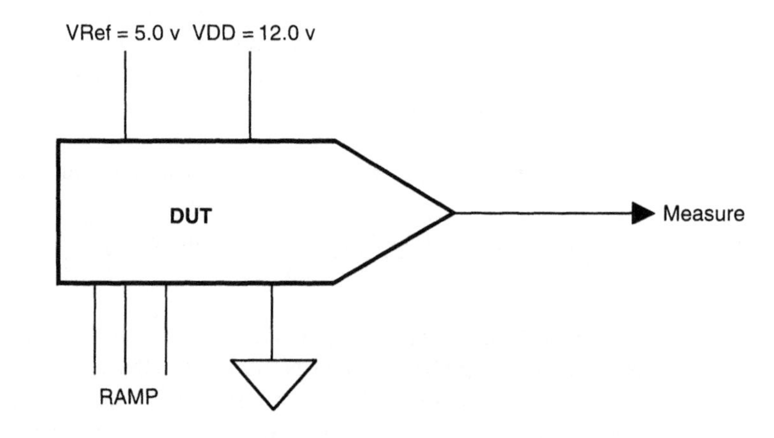

Figure 7.10

DAC Linearity Test Circuit

1. Apply power to the device power pins.
2. Apply the voltage reference level to the VREF pin.
3. Set the VIL and VIH level for the digital input pins.
4. Apply a ramp sequence of digital input codes.
5. Measure and record the analog output level for each code.
6. Calculate the DNL and INL from the analog output values.

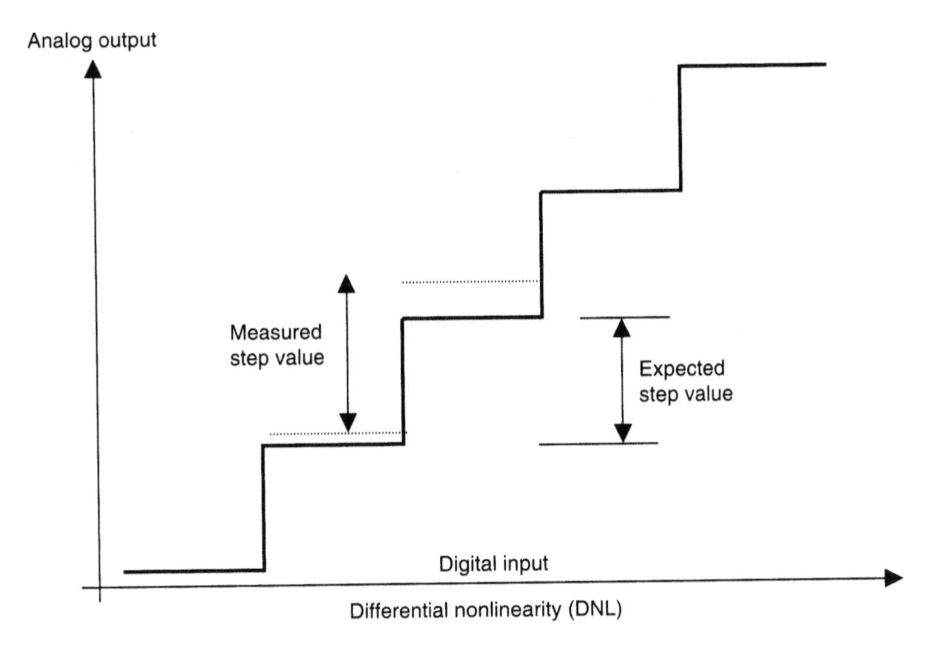

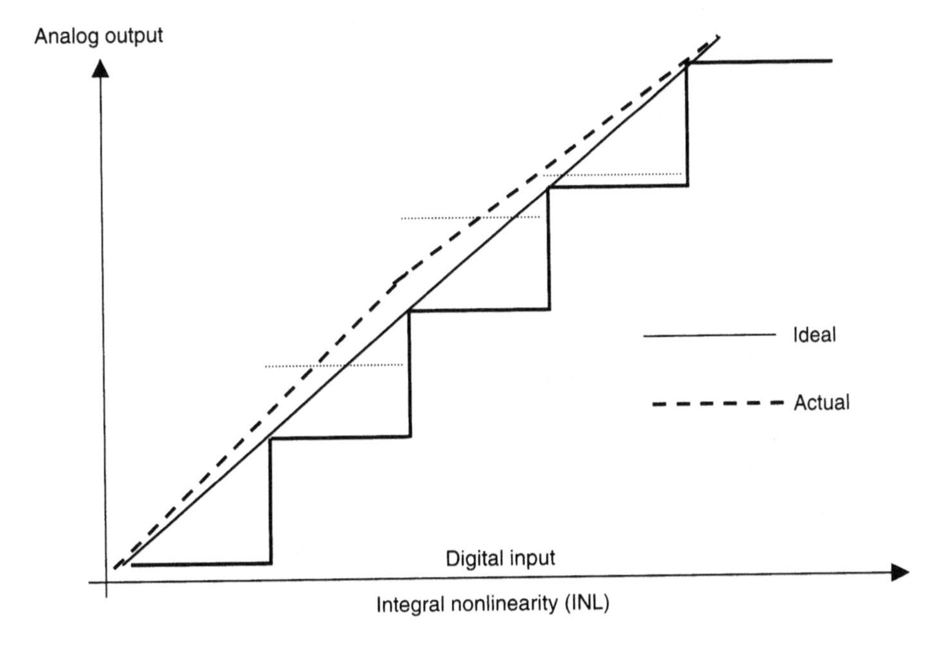

Figure 7.11

Differential and Integral Nonlinearity

7.4.1 Differential Linearity Tests

Let's look at a 4-bit DAC with a nominal full-scale range (REFERENCE LEVEL) of 10.0 volts. To determine the first endpoint, you drive the device with the smallest digital code, 00, and measure the voltage at 0.01 volts. For the next step in your test, you drive the device with a digital code of 15 and measure the output of the device at 9.985 volts.

Applying the device LSB equation,

$$\text{DeviceLSB} = \left(\frac{\text{Voltage}[2^n - 1] - \text{Voltage}[0]}{2^n - 1}\right)$$

for this device produces a device LSB value of 665 mV.

$$\text{DeviceLSB} = \left(\frac{9.985\text{ v} - 0.01\text{ v}}{15}\right) = 665\text{ mV}$$

To test DNL, each analog output step is measured and compared with the device LSB value.

Example for One Code Step: **(this would be repeated for each code)**

1. Driving the device with a digital code of 1100, the DUT output voltage is measured at 8.00 volts.
2. For the next step in the test, the device is driven with a digital code of 110, and output of the device is measured at 8.7 volts.
3. Calculate the value of the DNL for this bit.

   The DNL equation

$$\text{DNL} = \left(\frac{\text{V[i+1]} - \text{V[i]}}{\text{DeviceLSB}}\right) - 1.0$$

$$\text{DNL} = \left(\frac{8.7\text{ v} - 8.0\text{ v}}{665\text{ mV}}\right) - 1.0$$

   The DNL for this device, at this code, is therefore

$$\text{DNL} = \left(\frac{700\text{ mV}}{665\text{ mV}}\right) - 1.0 = 1.05 - 1.0 = 0.05\_\text{DLSB}$$

   That's not bad for a 4-bit DAC. Typically, the worst-case specification for DNL is +/–0.5 LSB, or half of a step.

7.4.2 Integral Linearity Tests

INL testing checks the overall "flatness" of the conversion range. An *ideal* DAC would have a straight line from the LSB value to the MSB value, with all of the intermediate codes in perfect alignment. An *actual* DAC will exhibit a curve from the ideal straight line, expressed as INL. Integral nonlinearity (INL) is a way of measuring the accumulated linearity over the entire range.

As with DNL testing, the first step is to measure the endpoints of the DAC, which are the analog output values corresponding to the smallest and largest digital input codes. From the measured endpoints, the device LSB is calculated as the reference step size for that device.

$$\text{DeviceLSB} = \left(\frac{\text{Voltage}[2^n - 1] - \text{Voltage}[0]}{2^n - 1} \right)$$

If the device is perfectly linear, the output level for each step should be a multiple of the device LSB step, referenced to the offset. The INL test calculates the difference between a given output level and the expected value, referenced to the device LSB.

$$\text{INL}[i] = \frac{\text{V}[i] - ((\text{DeviceLSB} \times i) + \text{V}[0])}{\text{DeviceLSB}}$$

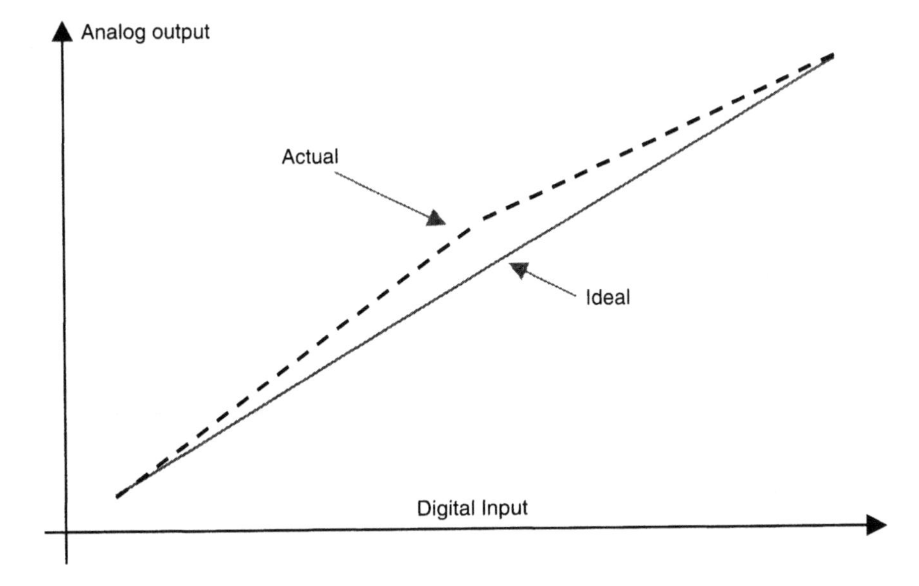

Figure 7.12

Integral Nonlinearity

Example INL test for one code step: **(this would be repeated for each code)**

What is the ideal voltage you would expect if you applied a digital code corresponding to the halfway point on a 4-bit DAC? (15 codes full scale = 0111 half point input code)

1. From the previous tests, the device LSB was calculated as 665 mV. The device LSB value multiplied by the test code (7) = 7 × 665 mV = 4.665.
2. The "start point" is the offset, or the output level for an "all-zeroes" code. Adding the offset voltage of 10 mV = 4.665 + 10 mV = 4.675, as the value of the expected output level with an input code of 0111.
3. Applying the digital input code of 0111 (7) produces an output value of 4.98 volts. Calculating the INL value for this bit, what is the difference between the measured value and the expected straight-line value, referenced to the device LSB?

$$\text{INL}[i] = \frac{V[i] - ((\text{DeviceLSB} \times i) + V[0])}{\text{DeviceLSB}}$$

$$\text{INL}[i] = \frac{4.97\ \text{v} - 4.675\ \text{v}}{665\ \text{mV}} = \frac{295\ \text{mV}}{665\ \text{mV}} = 0.44\ \text{DLSB}$$

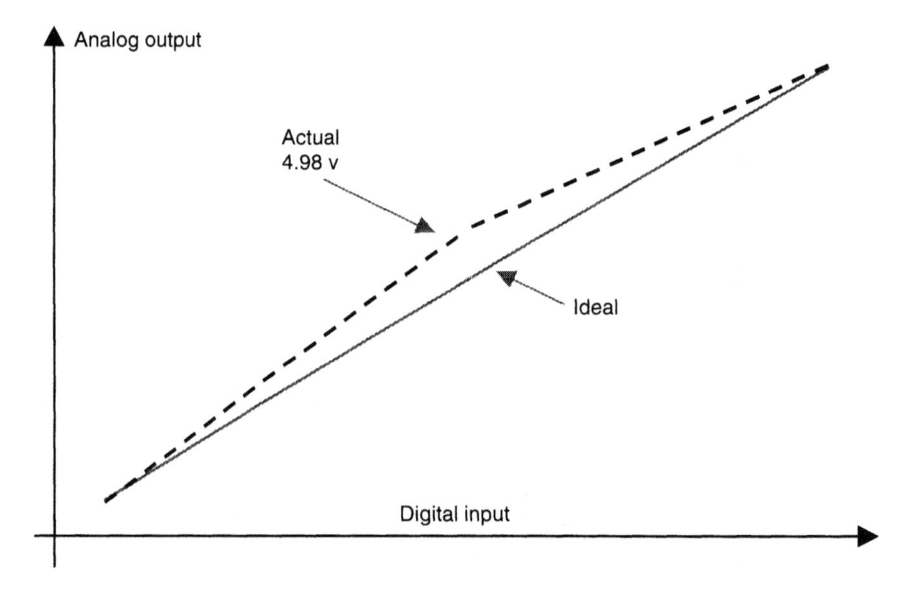

Figure 7.13

INL Test Application Example

If you're like me, you're probably thinking, "OK, I get the idea…but why do I have to measure each individual step twice, once for DNL and again for INL?" Well, as it turns out, you don't have to measure each step twice. Once you have collected the per-step measurement information for the DNL test, then you have already collected everything you need for the INL test. Just run an integral across the DNL data, which works like a running average. The maximum absolute value of the integral results will give you the worst-case INL.

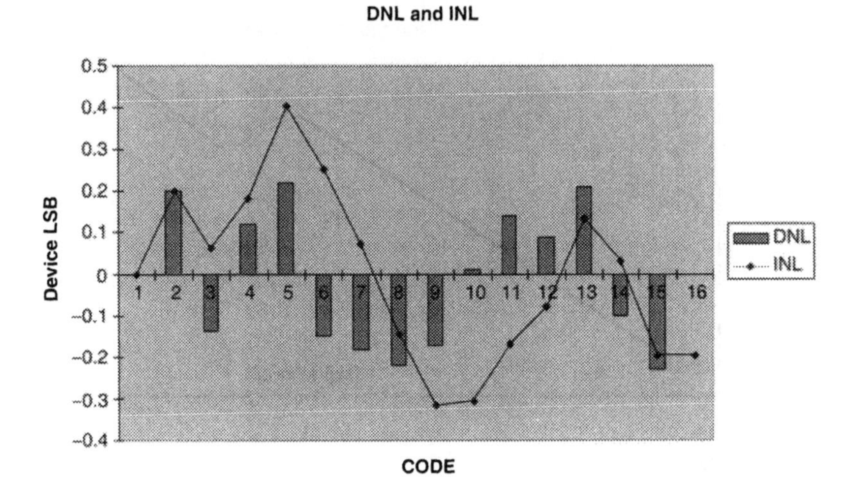

Figure 7.14

Calculating INL from the DNL Data Set

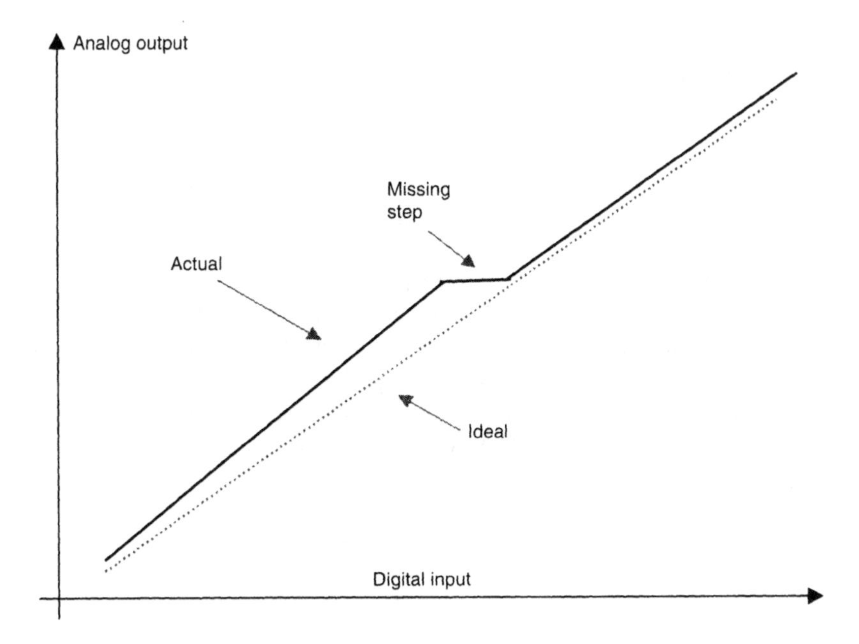

Figure 7.15

Missing Step Transfer Curve

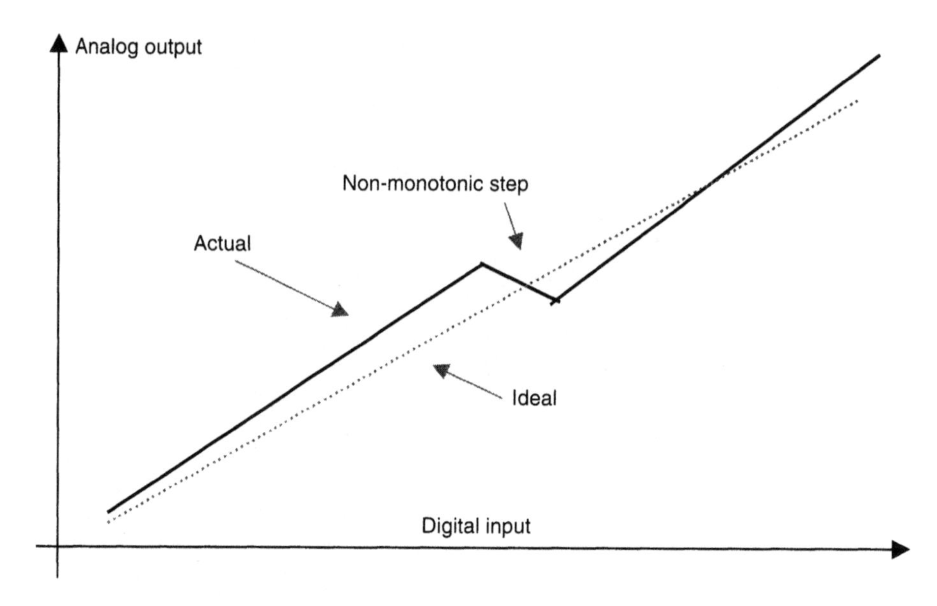

Figure 7.16

Non-Monotonic Transfer Curve

7.4.3 Missing Steps

A missing step in a DAC is sometimes called a missing code. Properly defined, a missing code is an error that can occur with an analog-to-digital converter (ADC). The output of a DAC will always be at some voltage level, whereas an ADC can have a flaw that prevents a certain code from occurring.

A DAC has a missing step if an increase in the digital code input does not result in an increase in the analog output. A missing code can occur if the Differential Nonlinearity Error (DNL) is more than 1/2 of an LSB. Testing for missing steps consists of checking that every increase in the digital input code causes a corresponding increase in the analog output level. No missing steps can be inferred by testing for DNL errors of less than 1/2 LSB value.

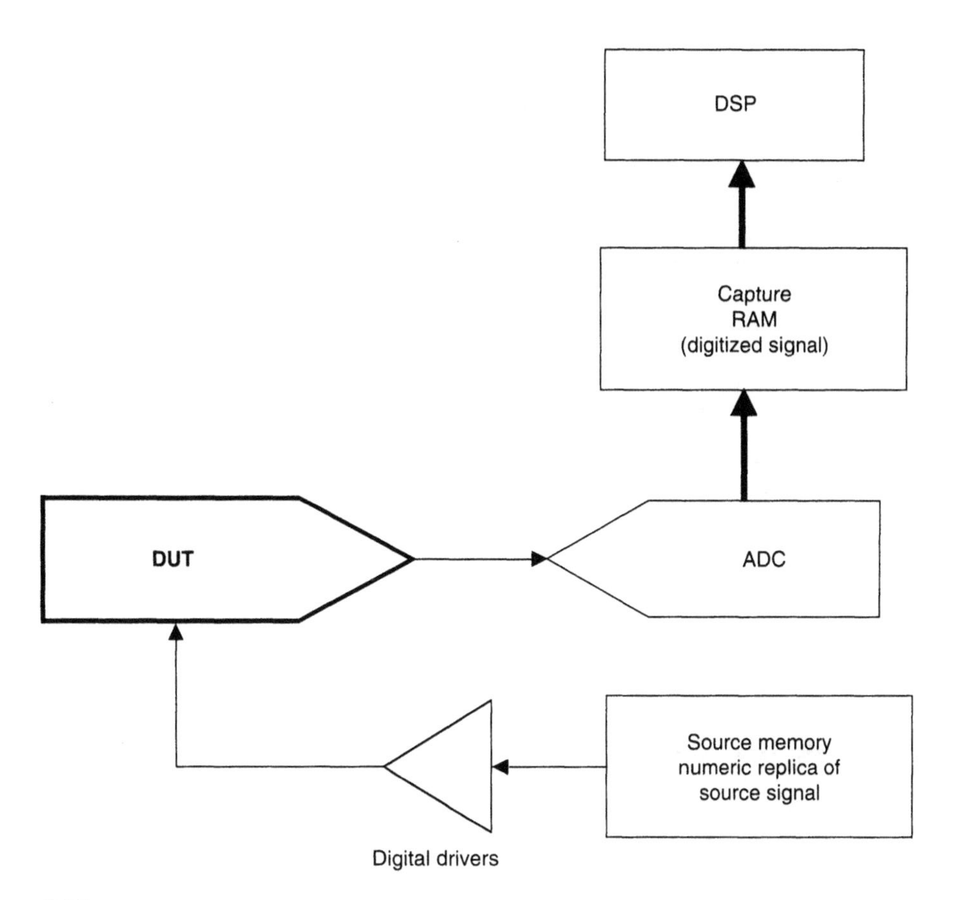

Figure 7.17

AC Test Setup

7.4.4 Monotonicity

Monotonicity. Wow, what a mouthful! Try to say that ten times in a row. A DAC is said to be monotonic if the transfer characteristic slope has the same sign over the entire range. A DAC is non-monotonic if an increase in the digital code input causes a decrease in the analog output value. Testing that a DAC is monotonic consists of checking that every increase in the digital input code causes a corresponding analog output level that is greater than, or equal to, the analog level from the previous code. Monotonic performance can be inferred by testing for DNL errors of less than 1 LSB value.

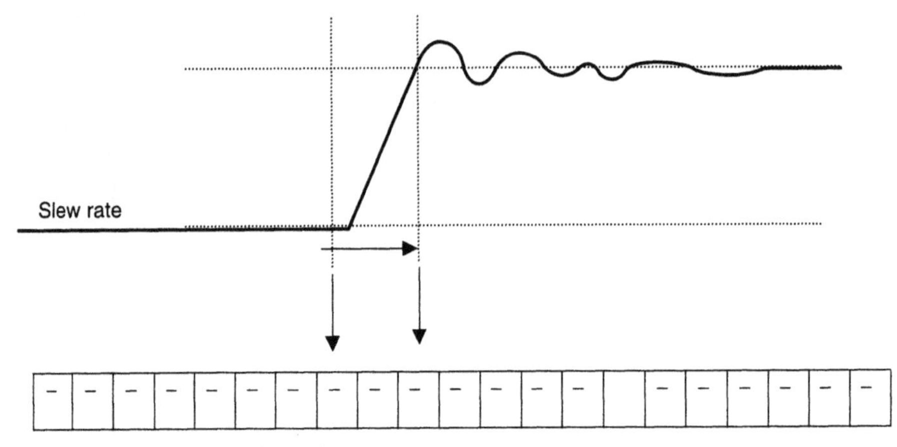

Figure 7.18

Slew Rate Test

7.5 AC Test Overview

Testing the AC performance of a DAC requires applying an AC signal or step function, capturing the DAC device output, and evaluating the response. A typical test setup for testing the AC performance of a DAC device uses the test system signal source to generate an AC signal, such as a sine wave. This "digital version" of the signal is routed to the digital pin electronics, which applies voltage level and timing formats. The DAC is driven with sequential digital vectors that cause the DAC analog output to generate the desired analog output.

As it turns out, not all DACs require the same type of AC testing. In fact, some device types do not require any AC testing at all because the end-use application is essentially static. Even for devices that do require AC testing, not all AC parameters make sense. For example, it doesn't make any sense to test for noise and distortion for a video DAC because a video DAC is designed to produce pixel luminance levels, and is not intended to reconstruct analog signals.

The test program must digitize the device output signal and then process the signal data with the test system's DSP. In the case of DC testing, the system requirements concern amplitude resolution. In the case of AC testing, the speed of the digitizer is of greater concern than in DC testing. Some test systems have a choice of digitizers, allowing the test engineer to choose between a high-accuracy digitizer for DC tests, and a high-speed digitizer for AC tests.

The theoretical minimum digitizer frequency is derived from Nyquist's theorem, which states that the digitizing sample rate must be at least twice the frequency of the input signal (two samples per signal cycle). In practice, it is more common to use a sample rate that is 8 to 16 times greater than the signal frequency, in order to provide the proper frequency resolution (fbase).

Depending on the intended application, some DAC devices may require testing to verify AC performance. Not all devices require all tests—the significant AC parameters are defined by the application.

7.5.1 Example AC Specifications

Slew Rate: The slope of the analog output signal across amplitude and time.

Settling Time: The elapsed time between the beginning of the analog output signal transition and the new analog output level.

Glitch Impulse Area: The amount of analog output amplitude variations across time.

Distortion: The ratio of periodic signal error amplitude to signal amplitude.

Signal-to-noise Ratio (SNR): The ratio of nonperiodic error amplitude to signal amplitude.

AC, or dynamic, testing of digital-to-analog converters requires time domain and frequency domain measurements. The test system must be capable of

- Synchronizing the digital input codes with the device under test.
- Measuring the time delay between the digital input and the analog output.
- Producing a sequence of digital codes that represent a dynamic analog signal.
- Capturing the dynamic analog signal produced by the device.
- Performing time domain and frequency domain measurements on the captured dynamic analog signal.

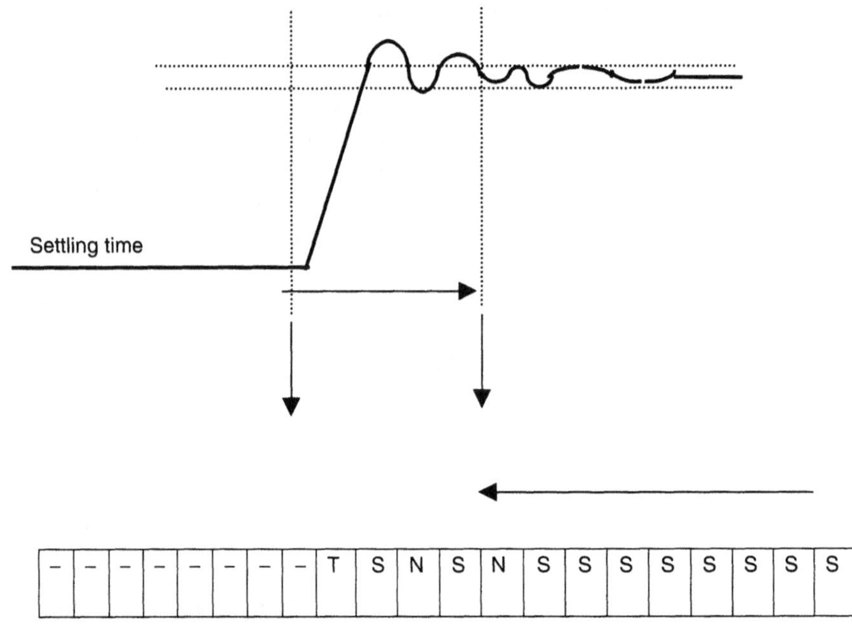

Figure 7.19

Settling Time Test

7.5.2 Slew Rate Test

To test slew rate, the DAC is driven with a step function digital input sequence, such as the minimum input code followed by the maximum input code. In response, the analog output of the DAC should swing from the minimum to the maximum value. The captured signal from the device output is analyzed by the DSP unit. The DSP calculates the slew rate by measuring the period between two thresholds of the signal slope.

Test Sequence

1. Apply power to the device power pins.
2. Apply the voltage reference level to the VREF pin.
3. Set the VIL and VIH level for the digital input pins.
4. Apply a step function sequence of digital input codes.
5. Capture the analog output response.
6. Use a DSP sequence to evaluate the time between the beginning of the slope transition to the upper slope threshold.

A first-order differentiation process is sometimes used to determine the location of the DUT output transition. One simple method for differentiation is to copy the captured array into a secondary array, with an offset of one sample. Subtracting the two arrays will produce a peak at the array element corresponding to the signal transition.

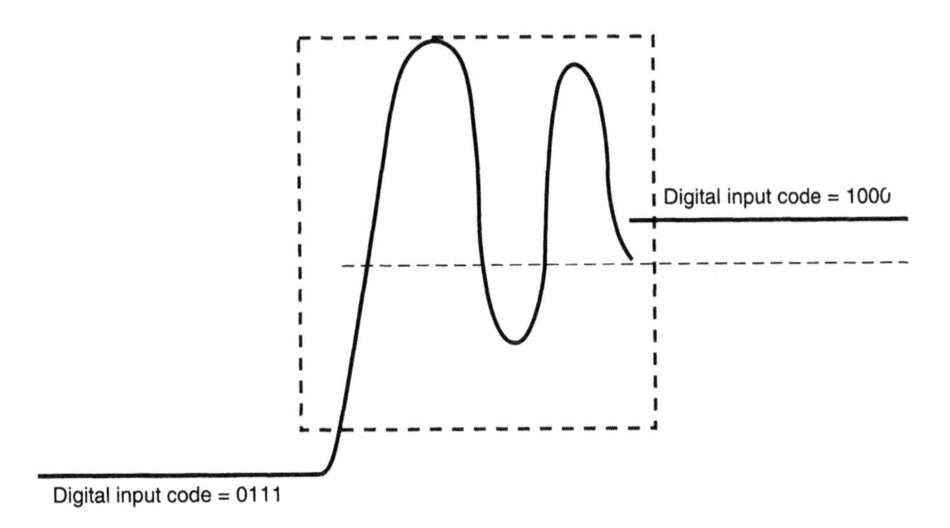

Figure 7.20

Glitch Impulse Area

7.5.3 Settling Time Test

Settling time of a DAC is usually defined as the time from beginning of the analog output transition until the DAC analog output settles to within a certain tolerance, typically 1/2 of an LSB. Settling time therefore expresses the device "propagation delay" in terms of analog performance. In order to properly check for any effects of ringing, the test is usually programmed to capture for a duration that is longer than the settling time specification. The captured waveform is analyzed "in reverse" to detect any out-of-boundary transitions. Evaluating settling time measures the transition from a +/−1/2 LSB band centered around the initial value, until the output settles within +/−1/2 LSB of the final value.

Test Sequence

1. Apply power to the device power pins.
2. Apply the voltage reference level to the VREF pin.
3. Set the VIL and VIH level for the digital input pins.
4. Apply a step function sequence of digital input codes.

5. Capture the analog output response.
6. Use a DSP sequence to evaluate the time between the initial analog output transition and the settled analog output value.

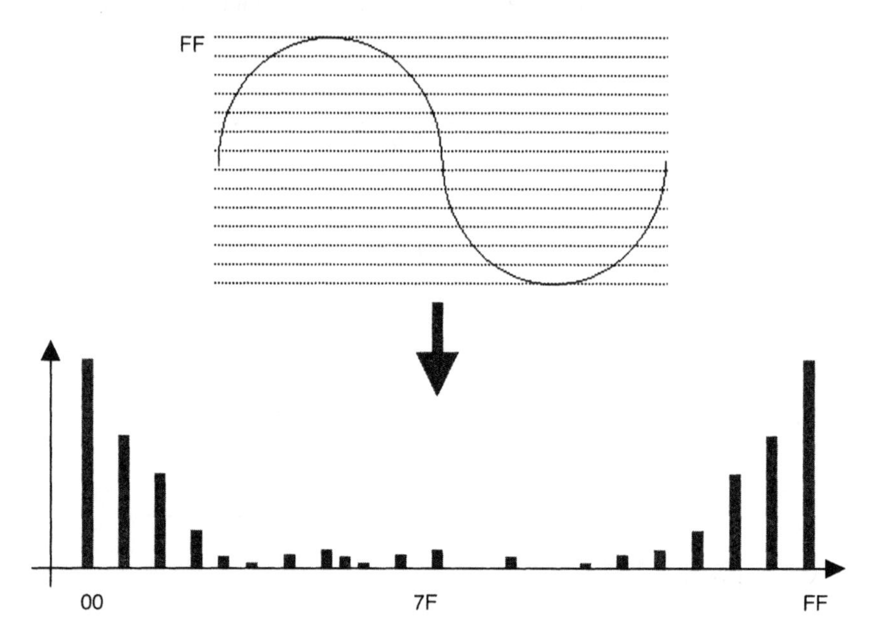

Figure 7.21

Sine Wave Histogram Data

The DSP analyzes the captured signal sample set "in reverse," from right to left—or backward in time. Beginning with the last element in the array, the DSP searches in reverse order for the first element that is not within the settled value range.

7.5.4 Glitch Impulse Area Test

Spurious transitions in a DAC analog output level are referred to as "glitches." Glitches are caused by timing skews in the internal digital logic and by unequal propagation delays through the DAC analog switches. Glitch testing is often performed at the transitions where all of the digital input bits are changing state—for example, at mid-scale. For example, a 4-bit DAC would have an input code of

0111	at just below mid-scale,
1000	at mid scale

The output level should change by a single LSB for this example code transition. However, internal to the device, all of the switches must change state. Timing skews and switch delays cause temporary glitches to occur in the analog output before reaching the new level. Because the practical effects of glitching depend on both the duration and amplitude of the spurious output, both the duration and amplitude of the glitch are measured together as "glitch area."

Some references describe glitch impulse area in terms of "pico-volts per second," which is a misnomer. The actual calculation is pico-volts TIMES seconds, or pV × S. Some DAC architectures are designed to minimize the effects of glitching. Sigma Delta DACs are inherently "glitchless," and DACs incorporating a sample and hold amplifier on the output effectively "latch" the analog output only after the output has settled.

Table 7.1

Harmonic Distortion Table

Frequency	Frequency Bin Location	Signal Voltage	Voltage Squared
750 Hz (fundamental)	5	985 mV	—
1500 Hz (2nd Harmonic)	10	2.1 mV	4.41 μV
2250 Hz (3rd Harmonic)	15	6.2 mV	38.4 μV
3000 Hz (4th Harmonic)	20	0.9 mV	0.81 μV

The DSP routine analyzes the captured signal transition and determines the peak-to-peak amplitude of the glitch, and the glitch duration. The glitch impulse area is calculated as amplitude multiplied by time, as a ratio of pico-volts times seconds.

7.6 Dynamic Performance Tests

The overall effects of nonlinearity (DNL and INL) together with glitch impulses across the entire range of digital codes is sometimes more easily expressed in terms of the DAC's ability to accurately produce a full-scale analog signal. Testing for AC signal integrity requires that the device be driven with a digital sine wave, which is synchronized with the DAC conversion rate to make sure a maximum number of codes are tested.

7.6.1 Sine Wave Histogram

Intuitively, it might seem that the input sample set should have a number of samples equal to the total number of code combinations. For example, testing a 12-bit DAC would *seem* to require an input pattern with a minimum of 4096 samples. However, this approach guarantees that some codes will never be executed. Consider a histogram of a single cycle sine wave. At the positive and negative peaks, the slope is not as steep as the slope at zero crossing. Grouping the sine wave data set according to the number of events per value will show a large number of events around the peak values, and a small number of events at zero crossing. For a sine wave, the zero crossing area produces a large amplitude change across a small period of time. The histogram of a sine wave is bi-modal, and is sometimes referred to as "bathtub histogram" because of the shape of the histogram distribution.

In order to generate samples at all possible codes, the input data set is designed to generate at least five cycles of the signal frequency.

Frequency	Bin Location	Frequency Voltage	Voltage Squared
DC	0	—	0—
150 Hz (Noise)	1	91 μV	8.28 nV
300 Hz (Noise)	2	43 μV	1.84 nV
450 Hz (Noise)	3	57 μV	3.25 nV
600 Hz (Noise)	4	29 μV	0.84 nV
750 Hz (Signal)	5	985 mV	—
900 Hz (Noise)	6	72 μV	5.18 nV
1050 Hz (Noise)	7	41 μV	1.68 nV
1200 Hz (Noise)	8	18 μV	0.32 nV
1350 Hz (Noise)	9	39 μV	1.52 nV
1500 Hz (2nd Harm)	10	2.1 mV	—
1650 Hz (Noise)	11	22 μV	0.48 nV
1800 Hz (Noise)	12	33 μV	1.08 nV
1950 Hz (Noise)	13	20 μV	0.40 nV
2100 Hz (Noise)	14	17 μV	0.29 nV
2250 Hz (3rd Harm)	15	6.2 mV	—
2400 Hz (Noise)	16	37 μV	1.37 nV
2550 Hz (Noise)	17	52 μV	2.70 nV
2700 Hz (Noise)	18	62 μV	3.84 nV
2850 Hz (Noise)	19	15 μV	0.22 nV
3000 Hz (4th Harm)	20	0.9 mV	—
3150 Hz (Noise)	21	34 μV	1.16 nV
...	...	...	1.27 nV
9450 Hz (Noise)	63	21 μV	0.44 nV
TOTAL		**738 μV**	**36.16 nV**

Table 7.2

Signal-to-Noise Table

7.6.2 Harmonic Distortion Test

To test for harmonic distortion, the output of the DAC must be captured and analyzed with the test system digitizer and processing resources. The DSP system is used to perform a fast Fourier transform to convert the captured data into frequency domain information. The frequency domain data generated by the FFT is analyzed by first measuring the amplitude of input signal frequency, which becomes the reference point for the harmonic content ratio. The amplitudes for the frequencies that are integer multiples of the signal frequency are measured and summed, and then the results are calculated as a percentage, or as a dB ratio.

Step One: Measure the energy value of the signal frequency (fundamental).

Step Two: Sum the harmonic energy,

$$\text{Total\_Harmonic\_Energy} = \sqrt{2\text{nd}^2 + 3\text{rd}^2 + 4\text{th}^2 \cdots}$$

where 2nd, 3rd, 4th ... are the frequency domain values for the integer multiples of the signal frequency (fundamental).

Step Three: Calculate the percentage,

$$\text{Total\_Harmonic\_Distortion} = \frac{\text{Total\_Harmonic\_Energy}}{\text{Fund}} \times 100$$

or for dB calculations,

$$\text{THD\_ratio} = 20 \times \log\left(\frac{\text{Fund}}{\text{Total\_Harmonic\_Energy}}\right)$$

Example

The DUT is programmed to generate 5 cycles of a 750-Hz sine wave. The data is captured at 19.2 kHz, with a total of 128 samples.

1. Calculate the base frequency.

$$\text{fbase} = \frac{\text{fs}}{\text{samples}} = \frac{19.2\ \text{kHz}}{128} = 150\ \text{Hz}$$

2. Determine the frequency bin location of the fundamental and harmonics.

$$\text{fundamental\_frequency\_bin} = \frac{\text{fi}}{\text{fbase}} = \frac{750\ \text{Hz}}{150\ \text{Hz}} = 5$$

$$\text{2ndHarmonic\_frequency\_bin} = \frac{\text{fi}}{\text{fbase}} = \frac{1500\ \text{Hz}}{150\ \text{Hz}} = 10$$

$$\text{3rdHarmonic\_frequency\_bin} = \frac{\text{fi}}{\text{fbase}} = \frac{2250\ \text{Hz}}{150\ \text{Hz}} = 15$$

Notice that the frequency bins of the signal harmonics are at integer multiples of the fundamental frequency bin.

Interpreting the FFT Results for Harmonic Distortion

By processing the raw output of the FFT with a polar conversion, the real and imaginary components can be mapped to magnitude values equivalent to the RMS voltage level of the signal as an indicator of the signal energy. Using the RMS voltage as a reference, the output of the FFT can be organized as table of energy values per frequency.

1. Find the harmonic energy as the algebraic sum of the harmonic signal energy.

$$\text{Distortion} = \sqrt{2\text{H}^2 + 3\text{H}^2 + 4\text{H}^2} = \sqrt{4.41\ \mu\text{V} + 38.4\ \mu\text{V} + 0.81\ \mu\text{V}}$$

$$= \sqrt{43.62\ \mu\text{V}} = 6.6\ \text{mV}$$

2. Calculate the dB ratio of the signal to the distortion figure.

$$-20 \times \log\left(\frac{\text{Fundamental}}{\text{Distortion}}\right) = -20 \times \log\left(\frac{985\ \text{mV}}{6.6\ \text{mV}}\right) = -20 \times \log(149.2)$$

$$= -43.47\ \text{dB}$$

or, as a percentage,

$$\text{THD} = \frac{\text{Distortion}}{\text{Fundamental}} \times 100 = \frac{6.6\ \text{mV}}{985\ \text{mV}} \times 100 = 0.67\%$$

7.6.3 Signal-to-Noise Tests

The same frequency domain data set that is used to determine harmonic distortion can also be processed to derive the signal-to-noise ratio (SNR). By convention, the classic signal-to-noise measurement does not include the harmonic energy, only the nonharmonic error components. Because the SNR test is a statistical measurement, a valid number of noise components must be processed in order to generate a valid result. The capture rate (fs) and sample size must be chosen to produce a statistically valid number of data points, and a suitable bandwidth corresponding to the SNR specification.

The signal-to-noise ratio (SNR) of a digital-to-analog converter is tested by driving the DUT with a digital replica of a sine wave, with a known frequency and amplitude. The analog output of the device is captured as digital data by the digitizer and analyzed in the frequency domain by use of an FFT algorithm. Energy other than the DC and signal frequencies is a digitizing error known as noise. Testing for the SNR processes the result of the FFT to remove the energy components due to the DC value, the signal energy, and the harmonic energy. The ratio of the signal amplitude to the noise level is expressed in terms of decibels (dB).

Step One: Measure the energy value of the signal frequency (fundamental).

Step Two: Sum the noise energy

$$\text{Noise\_Energy} = \sqrt{s1^2 + s2^2 + s3^2 + \cdots + sn^2}$$

Where s1, s2, s3, through s*n* are the frequency domain data points (frequency bins) that exclude the DC, fundamental, and harmonic signal components. A statistically valid number of noise samples, represented by the frequency bins values, is required.

Step Three: Calculate the db Ratio.

$$\text{SNR} = -20 \times \log\left(\frac{\text{Fund\_Energy}}{\text{Noise\_Energy}}\right)$$

Noise Test Using the FFT

Reviewing the example, the DUT is programmed to generate 5 cycles of a 750-Hz sine wave. The data is captured at 19.2 kHz, with a total of 128 samples. The FFT will generate a frequency domain data set of 64 samples, with a bandwidth of DC to 9.6 kHz (fs/2).

1. Noise total equals square root sum of squares:

$$\sqrt{38.16\ \text{nV}} = 190\ \mu\text{V}$$

2. SNR equals

$$-20 \times \log\left(\frac{S}{N}\right) = -20 \times \log\left(\frac{985\ \text{mV}}{190\ \mu\text{V}}\right) = -74.29\ \text{dB}$$

Chapter Review Questions

1. What is meant by INL?

2. What is meant by DNL?

3. Why would both an INL and a DNL test be required to test device linearity?

4. What is the calculated LSB step size (device LSB) for a 4-bit DAC with an offset of 200 mV and a full-scale output of 9.7 volts?

5. What is offset error? How is it tested?

6. What is gain error? How is it tested?

CHAPTER 8

TESTING ANALOG-TO-DIGITAL CONVERTERS

Fast is fine, but accuracy is everything.

—Wyatt Earp

8.1 Introduction

I used to say that testing an ADC was twice as difficult as testing a DAC. I was way off. It's a lot harder than that. Much of the terminology for ADC testing is similar to the terminology for DAC testing, but the methodology is much different. An ADC produces a set of digital codes that correspond to the analog input level. Like a DAC, testing the DC performance of an ADC consists largely of verifying a consistent and linear response. In practice, however, there is very little similarity between testing a DAC and testing an ADC. Because the digital output for an ADC is valid for a range of input values, ADC testing is more complex than testing a DAC. The methods used for DAC testing generally cannot be re-arranged and applied to testing ADC devices.

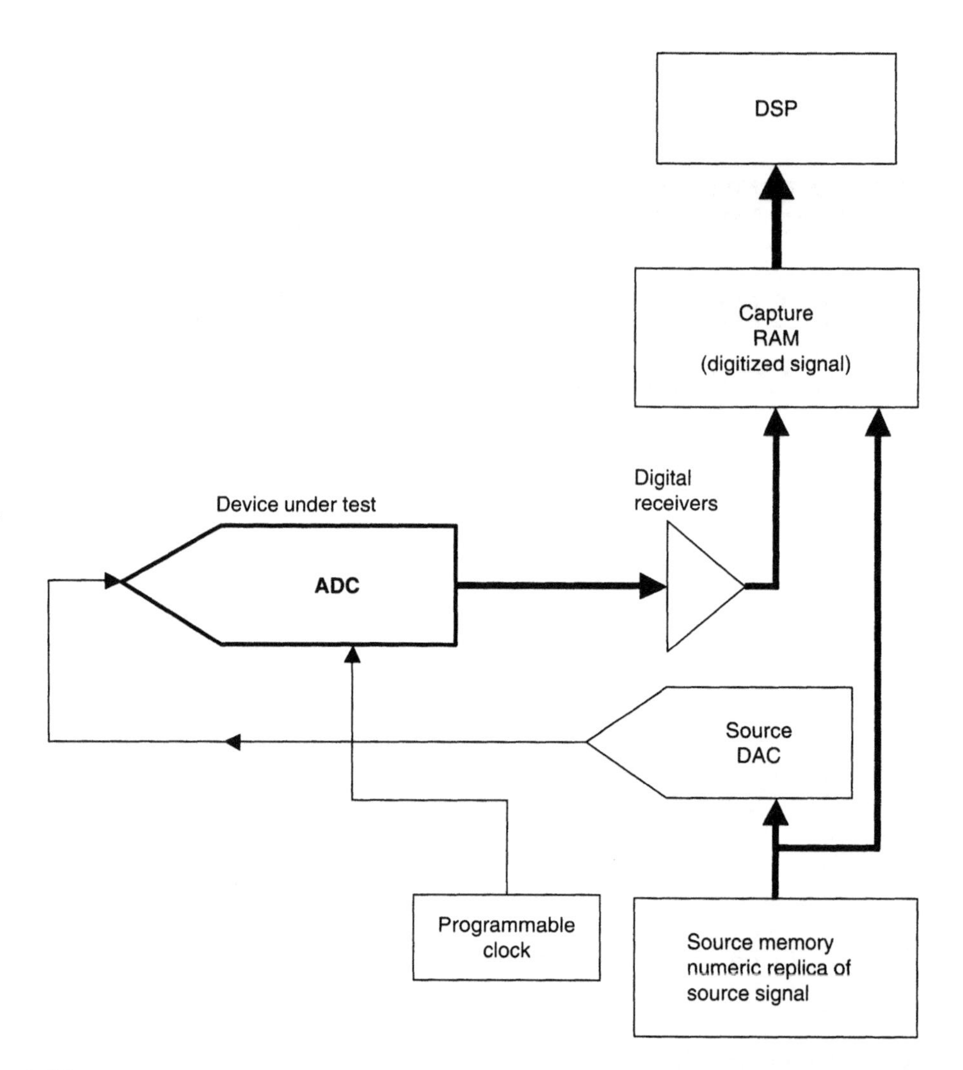

Figure 8.1

ADC Test Overview

An analog-to-digital converter is non-deterministic from the output to the input. If a test condition specifies a certain voltage on the input of a perfect ADC, one can predict the digital output code. However, if only the digital output code is known, there is no way to predict the exact input voltage—only its range can be predicted. It is not possible to correctly test an ADC by treating the device like a reverse DAC. Simply applying DC voltages and checking if the ADC has the correct response is not an adequate test method. In practice, testing for ADC linearity requires measurement of the voltage threshold that causes a code output change.

A typical test setup for testing the DC performance of an ADC device uses the test system signal source to generate an input voltage ramp via the ATE signal generator. The digital output codes generated by the ADC devices and the analog values corresponding to the output code transitions are then processed by the test system's digital signal processor (DSP). Whereas DAC testing requires a high-precision measurement system to verify the analog output, ADC testing requires a high-precision signal generator to produce high-accuracy analog input levels.

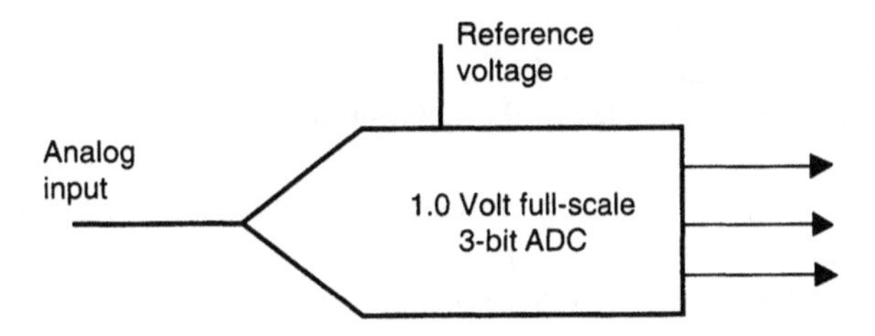

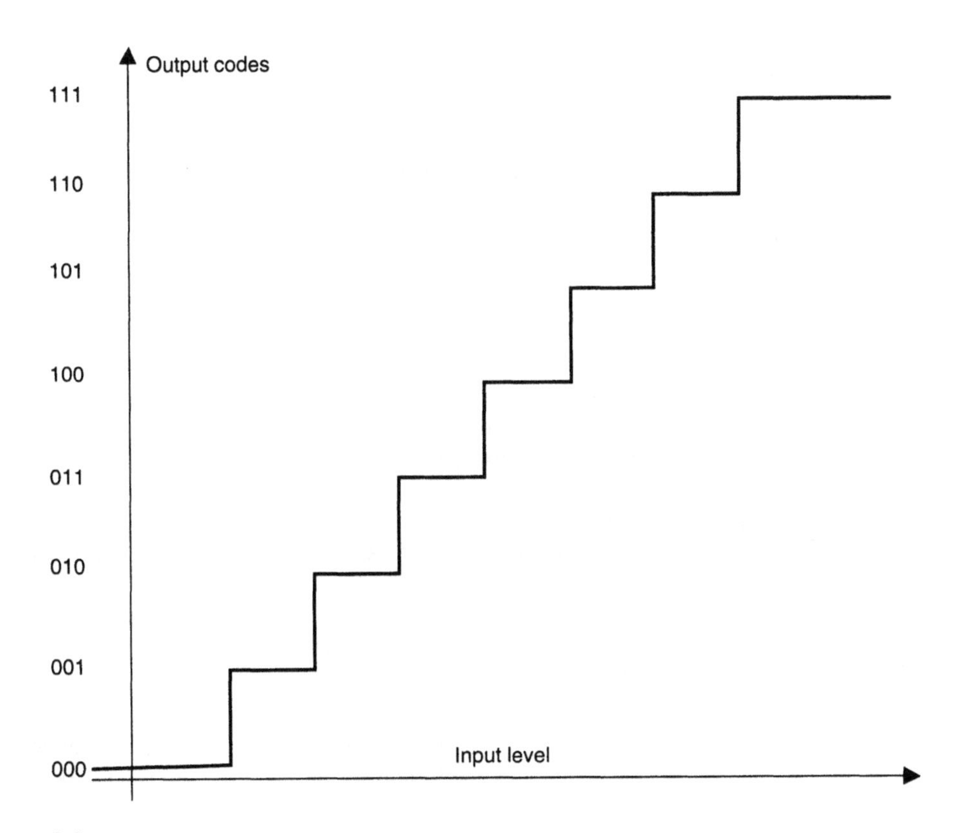

Figure 8.2

ADC Transfer Curve

8.2 ADC Overview

Let's look at a simple, general-purpose ADC to get some ideas about testing requirements. An analog-to-digital converter will generate a single output code for a range of input levels. The next highest code is output when the input level exceeds a given threshold. Evaluating the linearity of an ADC device measures the voltage range between code thresholds. The first output code transition occurs when the device changes from all zeroes to the next code step. The all-zero code does not have a corresponding input range, only an upper threshold. Likewise, the maximum output code identifies the threshold transition point, but there is not a corresponding range of values that correlates to the maximum output code.

Linearity tests for an analog-to-digital converter concern the relative size of each voltage step that causes a change in the output code. In this tutorial, the voltage thresholds that cause a change in the output code are called Code Boundaries (CB). Linearity tests reference either the code boundary or the calculated center between code boundaries (code center). Because the test concerns voltage spans corresponding to code changes, the measurements are made between, but not including, the two endpoints.

Referring to the illustration of the 3-bit ADC transfer diagram in Fig. 8.2, there are clearly 8 possible output codes, or 2^n, where n is the number of digital output bits. However, counting the step transitions across the x-axis indicates there are only 7 thresholds, or $2^n - 1$ code boundaries. When measuring the span between thresholds, a review of the transfer diagram shows there are only 6, or $2^n - 2$, equally spaced voltage spans corresponding to output code transitions.

Summary:

Number of unique output code values = 2^n.
Number of output steps = $2^n - 1$.
Number of input spans corresponding to an output code = $2^n - 2$.
where n equals the number of bits.

In this chapter, we will use the nomenclature CB[n], where CB stands for the Code Boundary input voltage threshold, or Code Boundary, and n refers to the corresponding output code.

8.3 DC Test Overview

As with DACs, there are two general categories of DC tests for ADC devices. The first category evaluates the device minimum and maximum input code boundaries, referenced to an absolute specification. ADC offset measures the variation of the input level causing the first output code transition. ADC gain measures the overall input span from the first code boundary to the last code boundary.

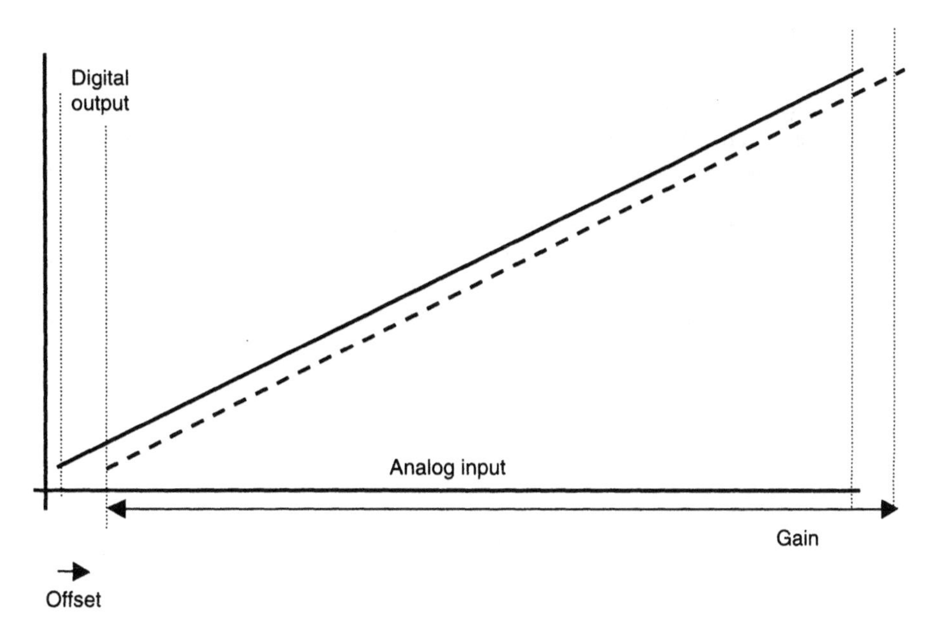

Figure 8.3

Offset and Gain Tests

The second category of DC tests evaluates the linearity by measuring the voltage range of the analog input steps that causes increments in the digital output code.

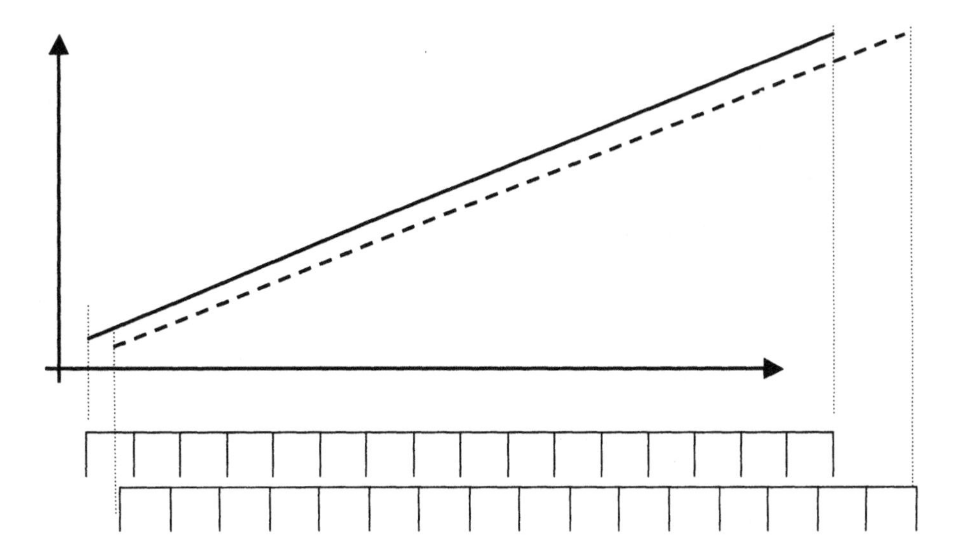

Figure 8.4

ADC Linearity Tests

8.3.1 Offset Measurement

ADC offset is the difference between the ideal and actual analog input values that cause a transition from "zero code" digital output to the next code increment. The ideal "zero code" transition is calculated as a fraction of the reference voltage range, or Full-Scale Range (FSR). Offset is also known as the "Zero Code Error."

Using our 1-volt, 3-bit converter as an example, the full-scale range would be 1 volt—that is, the reference voltage level. ADC devices are usually designed so that the first code transition should occur at one-half of the nominal step size, or LSB. The nominal or ideal step size is calculated by dividing the FSR by the number of steps (2^n). The ideal first step Code Boundary (CB) for this device should therefore be 1/2 of .125 volts, or 62.5 mV.

$$\text{Ideal\_LSB} = \frac{\text{FSR}}{2^n} = \frac{1.0 \text{ volts}}{8} = 0.125 \text{ volts}$$

$$\text{Ideal\_First\_Threshold} = \frac{\text{Ideal\_LSB}}{2} = \frac{0.125 \text{ v}}{2} = 62.5 \text{ mV}$$

If the measured first code transition threshold, CB[1], was measured at 72.5 mV, the difference between the ideal and actual indicates an offset error of 10 mV.

$$\text{Offset} = \frac{\text{CB}[1] - \left[0.5 \times \left(\frac{\text{FSR}}{2^n}\right)\right]}{\frac{\text{FSR}}{2^n}}$$

1. Calculate the ideal LSB voltage step as $\frac{\text{FSR}}{2^n}$.
2. Calculate the offset error as = CB[1] – 1/2 of an Ideal LSB.
3. Divide the error by the Ideal LSB to derive the fractional LSB offset value.

Some devices are designed so the first threshold is equal to 1 LSB, in which case the offset is referenced to FSR/2^n.

Offset Test Example

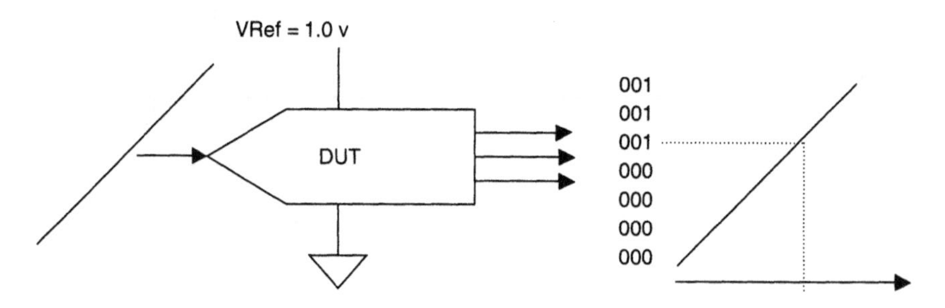

Figure 8.5

Offset Test Setup

1. Apply power to the device power pins.
2. Apply the voltage and adjust the reference level to the VREF pin.
3. Adjust the input voltage until the output code changes from 000 to 001.

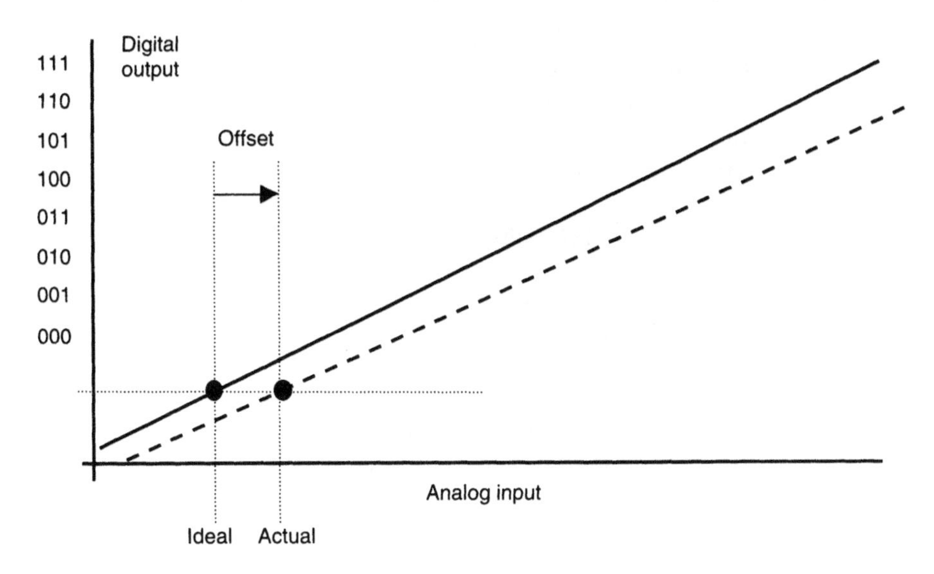

Figure 8.6

Offset Measurement

8.3.2 Gain Measurement

ADC gain is the difference between the ideal and actual span of analog input values corresponding to digital output codes. Because the offset shifts the entire transfer function, it is subtracted from the maximum code threshold level in order to normalize the reference. The total number of analog steps is $2^n - 2$, so the ideal span is equal to the full-scale range minus 2 ideal LSB steps. The full equation for the ideal span is FSR – (2 × LSB). The LSB is equal to FSR/2^n, so the ideal input span can be calculated as

$$\text{Ideal\_Span} = \text{FSR} \times \left(\frac{2^n - 2}{2^n} \right)$$

Example

1. $\text{Ideal\_Span} = 1\text{v} \times \left(\frac{6}{8} \right) = 0.75 \text{ volts}$
2. Measured span = Last code – First code

Measured last code transition = 0.875 volts
Measured first code transition = 72.5 mV
Measured span = 0.875 − 72.5 mV = 0.802 volts

3. Gain Error = Measured span/Ideal span

Measured span = 0.802 volts
Ideal span = 0.75 volts
Measured/Ideal = 0.802/0.75 = 1.07
1.07 − 1.0 = 0.07
0.07 × 100 = 7% Gain Error

Gain Error (in percentage of FSR)

$$\text{Gain Error} = \left(\frac{\text{Actual}}{\text{Ideal}} - 1\right) \times 100 = \%$$

$$\text{Gain Error} = \left[\left(\frac{CB[2^n - 1] - CB[1]}{FSR \times \left(\frac{2^n - 2}{2^n}\right)}\right) - 1\right] \times 100$$

Gain Test Example

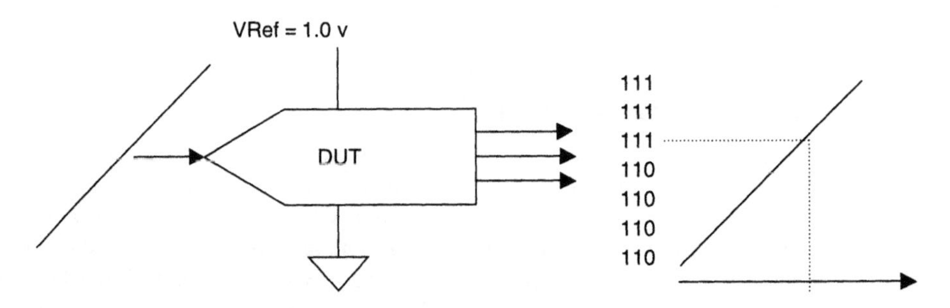

Figure 8.7

Gain Test Setup

1. Apply power to the device power pins.
2. Apply the voltage reference level to the VREF pin.
3. Adjust the input voltage until the output code changes from 110 to 111 to determine the $CB(2^n)$ value.
4. Subtract the first threshold value CB[1] from the last threshold value CB[111] to determine the DUT input span.
5. Compare the actual span with the ideal span as an error percentage.

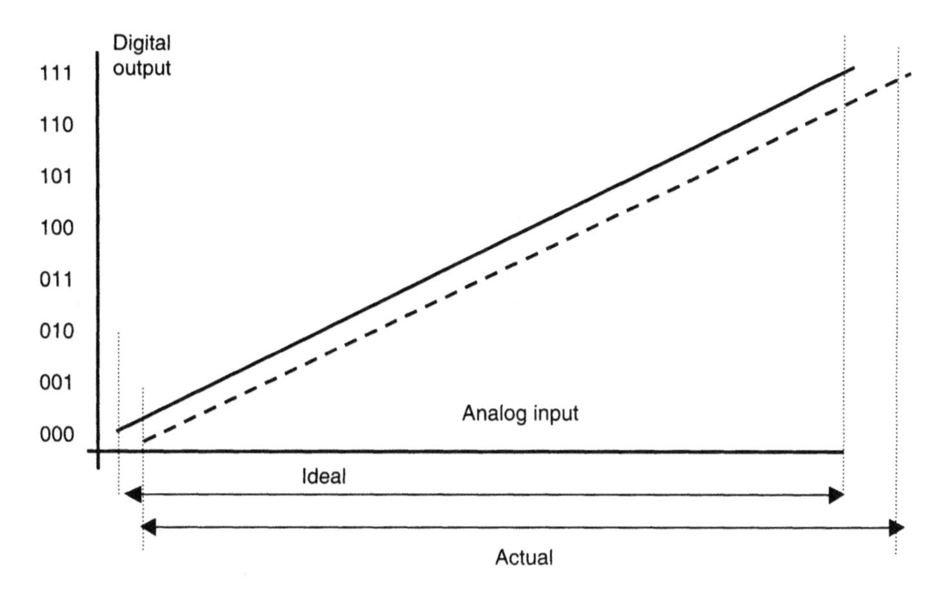

Figure 8.8

Gain Test Measurement

8.4 Linearity Test Overview

Testing ADC device linearity evaluates the device in terms of the analog input steps corresponding to increments in the output digital code. The analog input steps correspond to the difference between two adjacent code boundary (CB) values. Ideally, each increment in the analog input value causing a change in the digital output code would be exactly the same range. In an actual device, the analog step size varies. The linearity of the transfer function is referenced to a calculated device step size (LSB).

The device LSB step is calculated by dividing the total span of the ADC by the number of corresponding analog input transition steps, or code transitions. The first code transition is not "all zeroes," but actually the transition between all zeroes and the next increment. Therefore, the device LSB value is calculated referenced to the total number of codes, less the first code. Because the total number of codes transitions = $2^n - 1$, the total number of code spans is $2^n - 2$.

$$\text{Device LSB} = \frac{\text{CB}[2^n - 1] - \text{CB}[1]}{2^n - 2}$$

Differential Nonlinearity (DNL) is the difference between each analog increment step and the calculated LSB (device LSB) increment. DNL is also described as DLE, Differential Linearity Error.

$$\mathrm{DNL} = \left(\frac{\mathrm{CB}[i+1] - \mathrm{CB}[i]}{\text{Device LSB}} \right) - 1.0$$

Integral Nonlinearity (INL) is the worst-case variation in any of the code boundaries with respect to an ideal straight line drawn through the endpoints. INL is also sometimes defined in comparison to a "best fit" straight line. INL is also described as ILE, Integral Linearity Error.

$$\mathrm{INL}[i] = \frac{\mathrm{CB}[i] - (\text{Device\_LSB} \times (i-1) - \mathrm{CB}[1])}{\text{Device\_LSB}}$$

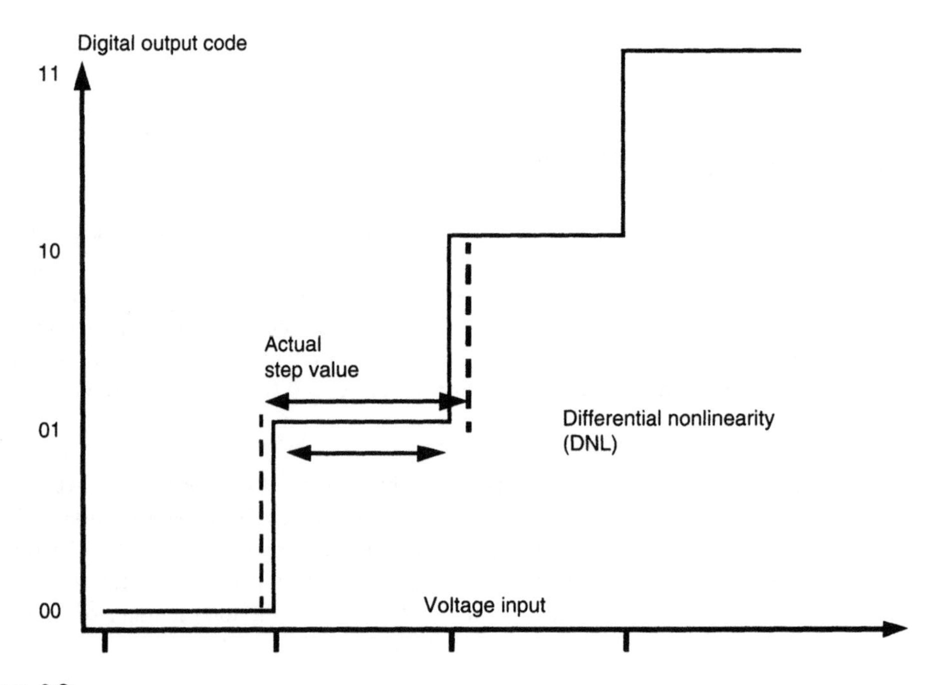

Figure 8.9

ADC Differential Nonlinearity

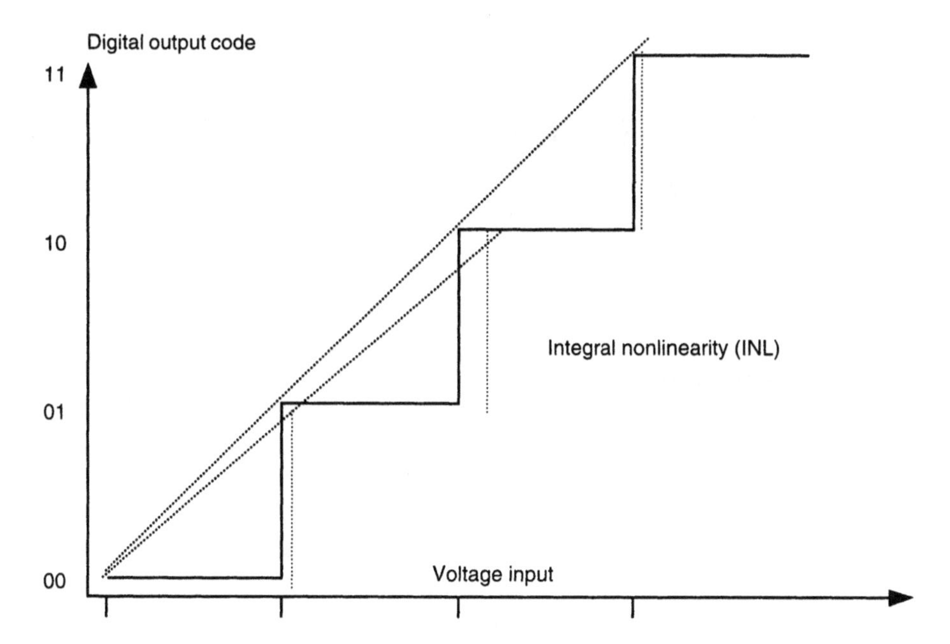

Figure 8.10

ADC Integral Nonlinearity

8.4.1 Differential Linearity

Example Device: 4-bit ADC with a 10.0 reference voltage

1. Calculate the actual device LSB code boundary width.

   The transition from 0000 to 0001 occurs at 0.3 volts, which is CB[1]. The transition from 1110 to 1111 occurs at 9.9 volts, which is CB[15].

   The device LSB is therefore

$$\text{Device LSB} = \frac{CB[2^n - 1] - CB[1]}{2^n - 2} = \frac{9.9\text{ v} - 0.3\text{ v}}{14} = \frac{9.6\text{ v}}{14} = 686\text{ mV}$$

2. Driving the device with a ramp, the input level that causes the output digital code to change from 1001 to 1010 is measured at 6.1 volts.
3. For the next step in your test, the voltage input is increased until the device responds with a digital code of 1011. The input voltage causing the output code change is measured at 6.7 volts. The voltage increment for this step is therefore 600 mV (6.7 volts – 6.1 volts = 600 mV).

4. Calculate the value of the DNL for this bit, as a fraction of a device LSB.

$$\text{DNL} = \left(\frac{\text{CB}[i+1]-\text{CB}[i]}{\text{Device LSB}}\right) - 1.0 = \left(\frac{6.7\text{ v}-6.1\text{ v}}{685\text{ mV}}\right) - 1.0$$

$$= \left(\frac{600\text{ mV}}{685\text{ mV}}\right) - 1.0 = 0.875 - 1.0 = -0.12$$

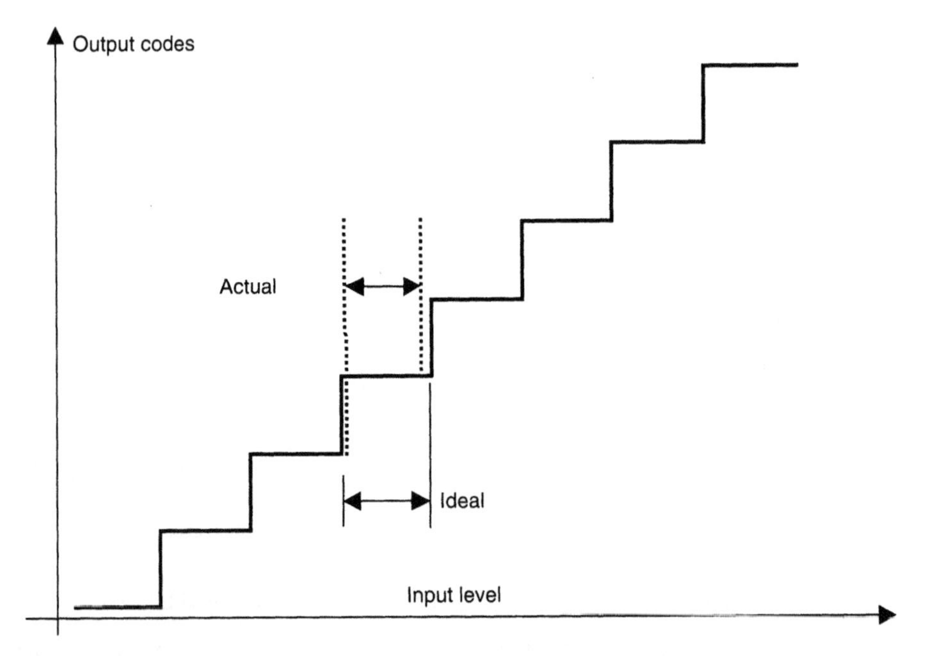

Figure 8.11

Differential Linearity Test Sequence

8.4.2 Integral Linearity

INL testing checks the overall "flatness" of the conversion range. An ideal ADC would have a straight line from the LSB value to the MSB value, with all of the intermediate codes in perfect alignment. An actual ADC will exhibit a curve from the ideal straight line, expressed as INL. From the previous example, we know that the voltage threshold for the first code transition is 0.3 volts. The voltage threshold for the last code transition is 9.9 volts. The linearity span is therefore 9.6 volts, which is divided by the remaining number of codes to give us the device LSB step-size of 665 mV. Remember that we are evaluating linearity based on the two endpoints derived from the offset and gain tests. When determining the "ideal straight line" of code boundaries, the calculations are referenced to the first code transition.

Example

1. The ideal straight-line intersection is calculated by multiplying the device LSB by the code value −1, referenced to the first code transition threshold, CB[1].

   $$\text{Ideal\_value} = (\text{Device\_LSB} \times (i-1)) + \text{CB}[1]$$

   For the example device, the device LSB was 665 mV, and the first code transition was 300 mV. For a perfectly linear device, we can predict that the threshold that generates the code of 0111 would be equal to the first code threshold, 300 mV, plus six additional steps of 665 mV each.

   $$\text{Ideal\_value} = (\text{Device\_LSB} \times (i-1)) + \text{CB}[1]$$

   $$\downarrow$$

   $$(665\text{ mV} \times 6) + 300\text{ mV} = 4.29\text{ v}$$

2. The actual input voltage level that causes the output code to change from 0110 to 0111 is measured at 4.6 volts.
3. The difference between the expected or ideal threshold value and the actual threshold value is a function of the device integral nonlinearity.

$$\text{INL}[i] = \frac{\text{CB}[i] - (\text{Device\_LSB} \times (i-1) + \text{CB}[1])}{\text{Device\_LSB}}$$

$$\downarrow$$

$$\text{INL}[i] = \frac{\text{Actual\_Threshold} - (\text{Ideal\_Threshold})}{\text{Device\_LSB}}$$

$$\downarrow$$

$$\text{INL}[i] = \frac{4.6\text{ v} - 4.29\text{ v}}{665\text{ mV}} = 0.466\_\text{DLSB}$$

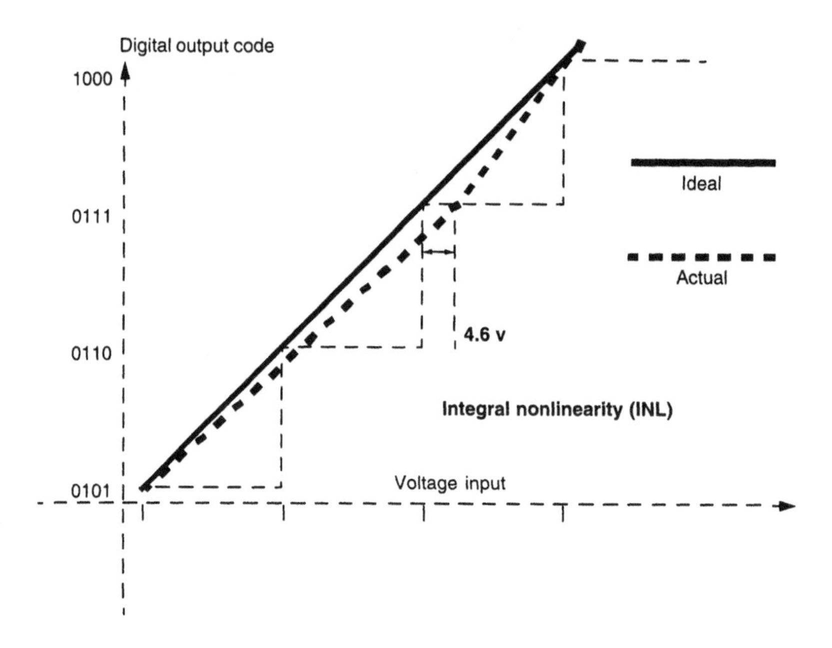

Figure 8.12

Integral Linearity Test Sequence

As with DACs, in practice the INL is derived from the collected DNL data. Performing an integral calculation on the DNL data set produces a "running average" that corresponds directly to the actual INL for each code. The maximum absolute value of the integral results is the same as the worst-case INL error.

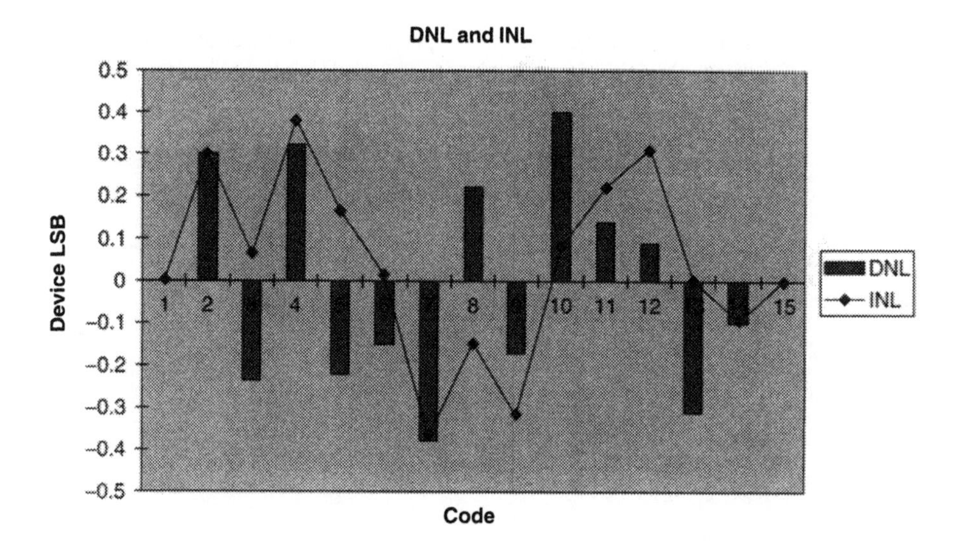

Figure 8.13

DNL Data Set with INL Integral

8.5 Missing Codes

An ADC is said to have a missing code if the digital output codes have a gap; that is, the missing code is an output value that is never generated. A missing code can occur if there is no analog input value that can generate a specific code. If a bit combination has a *positive* DNL error of one-half of an LSB, and the subsequent bit combination has a *negative* DNL error one-half of an LSB, the overall effect will be a missing code. No missing codes can be inferred by testing for DNL errors of less than 1/2 LSB value.

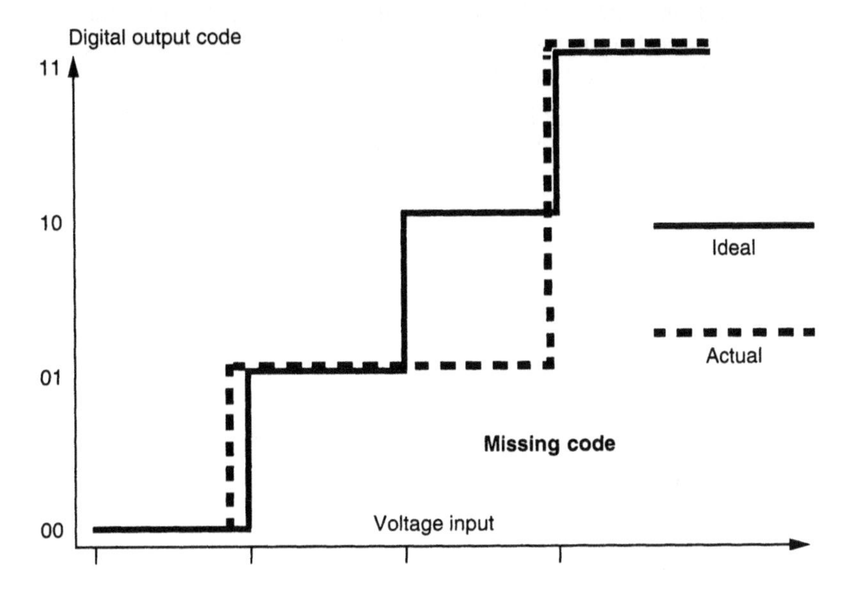

Figure 8.14

Missing Codes

Non-monotonicity may occur with digital-to-analog converters, but could only theoretically occur with an ADC device. The error would indicate a negative code width. Under dynamic conditions, an ADC can appear to be non-monotonic because of the variations in the code boundary thresholds.

8.6 The Histogram Test Method

The power of digital signal processing can be applied to testing the DC linearity for ADC devices by use of a histogram. A histogram is a data structure that organizes data according to the number of occurrences of a value. The output of the histogram algorithm can be visualized as a bar chart displaying the number of events corresponding to a measured value. A histogram test typically drives the input of the analog-to-digital converter with a ramp that begins at the device negative full scale and extends to the positive full scale. The ramp is synchronized with the device clock so that a given number of conversions takes place for each consecutive voltage span corresponding to an output code.

Example

A 12-bit ADC DUT features a device LSB of 2.4 mV. The ATE system is programmed so that each time the DUT is clocked, the signal source increments the DUT input level by 150 μV. When the input level reaches 3.6 mV, the DUT begins to generate an output code of ×02. The DUT continues to generate this output code until the input level reaches 6.0 mV, when it begins to generate an output code of ×03.

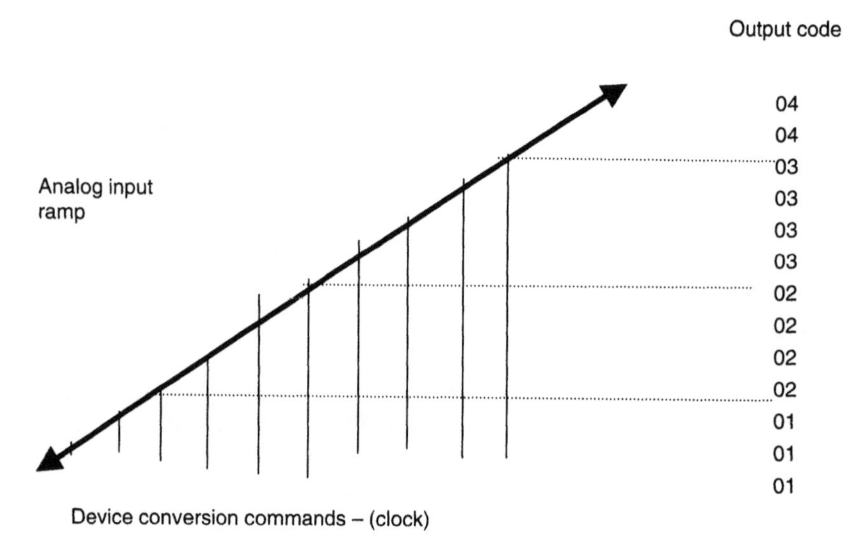

Figure 8.15

Collecting Output Code Values for the Histogram Test Method

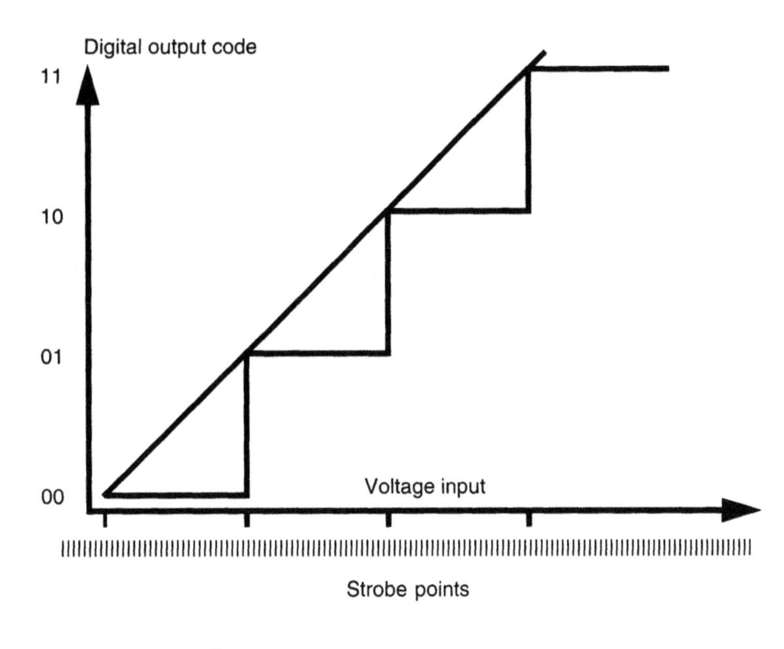

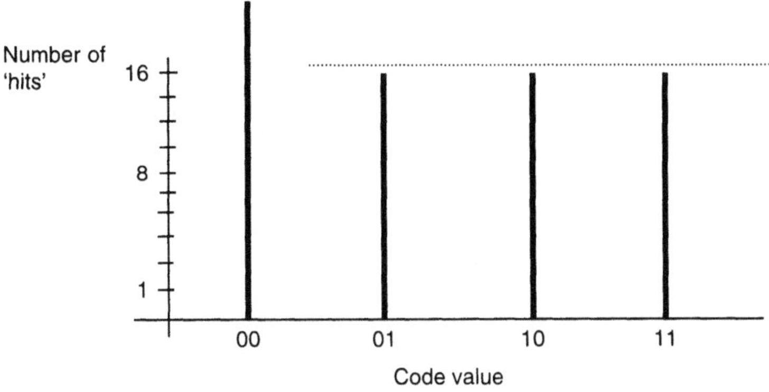

Figure 8.16

Histogram Data and DNL

If the DUT features equally spaced analog input spans corresponding to a change in the output code, each code will be generated an equal number of times. Hey, it could happen, right? For example, if the input voltage level is increased by $1/16^{th}$ of a device LSB for each conversion, then each code will be generated 16 times. Any variation in the code width will cause a variation in the number of occurrences of a given code, which corresponds directly to the DNL error. The number of events for the first and last codes per code is meaningless, because the first and last codes do not have a defined code width.

8.6.1 Events per Code

If the DUT has an input voltage span that is greater than or less than the device LSB, then the number of codes generated for that particular span will either be greater or lesser than expected. In this example, the output code transition from ×01 to ×10 did not occur until the input voltage level was above the expected threshold. As a result, the DUT continued to generate additional output code values of ×01. There is a direct correlation between the number of events in the histogram, and the amount of DNL error for that bit.

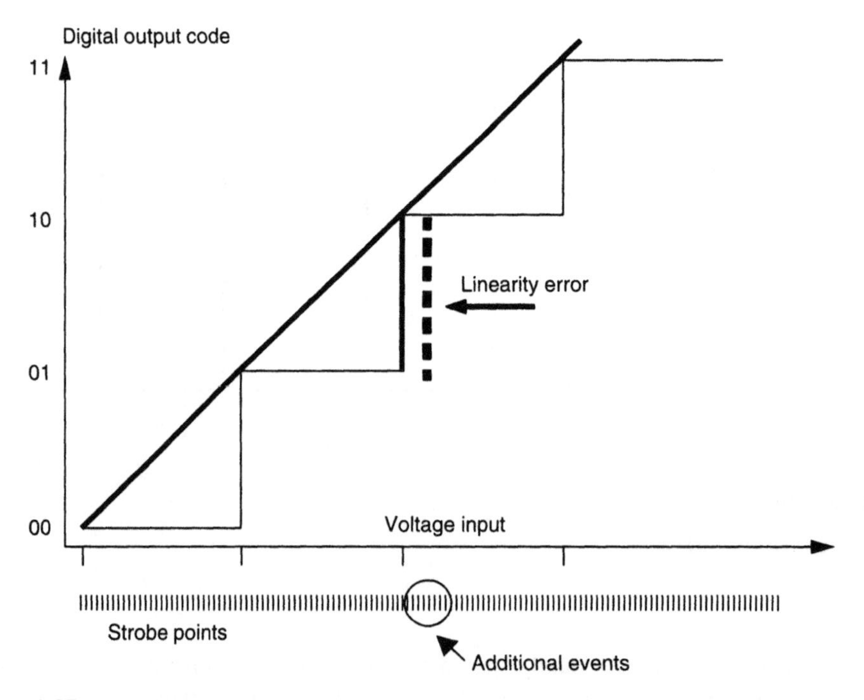

Figure 8.17

Code Events

8.6.2 Weighted Sine Wave Histogram

There are several different input waveforms that are commonly used for the histogram test method. Because the ADC code boundaries are subject to variation due to noise and hysteresis, a dual slope ramp, or triangle wave, often provides a more thorough test than a single full-scale ramp.

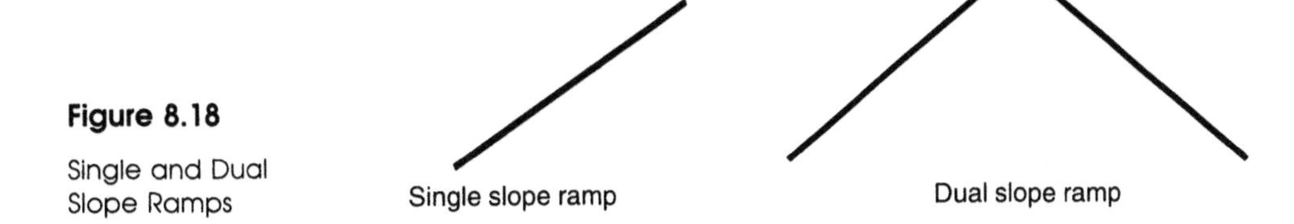

Figure 8.18

Single and Dual Slope Ramps

For devices intended for dynamic signal acquisition, a weighted sine wave histogram is sometimes the preferred approach. Instead of a ramp, the input signal is a multiple-cycle sine wave signal. The output data set is mathematically corrected, or weighted, to correct for the unequal histogram distribution of a sine wave. The unprocessed histogram of an ADC output sequence from a sine wave input will exhibit a "bathtub" distribution, along with a non-deterministic number of events for the lowest and highest code values. Correcting the sine wave histogram consists of removing the undefined endpoint values, and then scaling the histogram values to correct for the sine wave distribution.

$$\text{Ideal\_Events[code]} = \frac{\text{samples}}{\Pi} \times \left[\sin\left(\frac{\text{code} + 1 - 2^{n-1} - \text{offset}}{\text{CB}[2^n - 1]} \right) - \sin\left(\frac{\text{code} - 2^{n-1} - \text{offset}}{\text{CB}[2^n - 1]} \right) \right]$$

The actual number of events per code from the histogram data set is divided by the ideal number of events for that code to produce a normalized histogram.

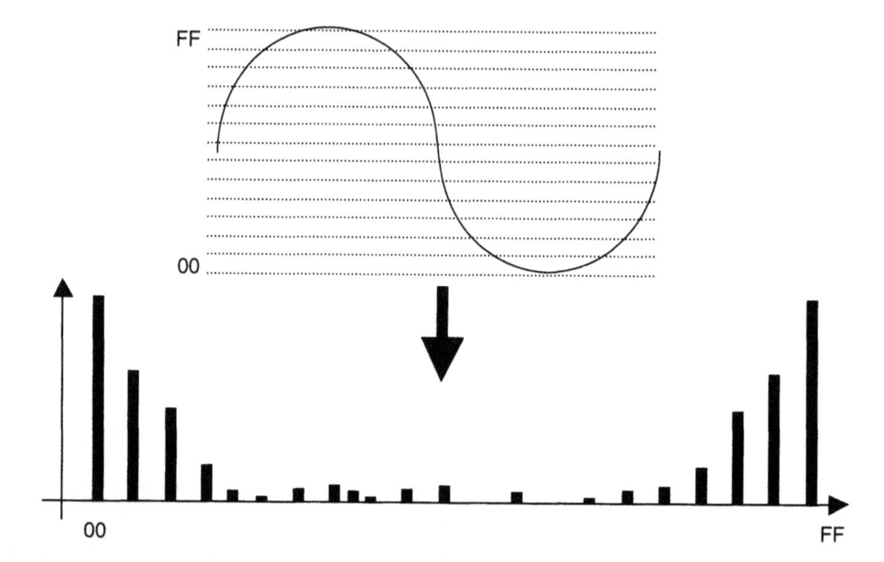

Figure 8.19

Sine Wave Histogram

8.6.3 The Segmented Ramp

A hardware-based signal generation technique called the "segmented ramp" is sometimes used in conjunction with the histogram test method. In order to generate a precise ramp for high-resolution ADC devices, the signal source is programmed to generate a sequence of ramps, each referenced to a different DC base level. The analog input of the device is driven with a very precise ramp, which is actually composed of several smaller ramps in succession. Consider the following example, in Fig. 8.20, using a 14-bit signal source.

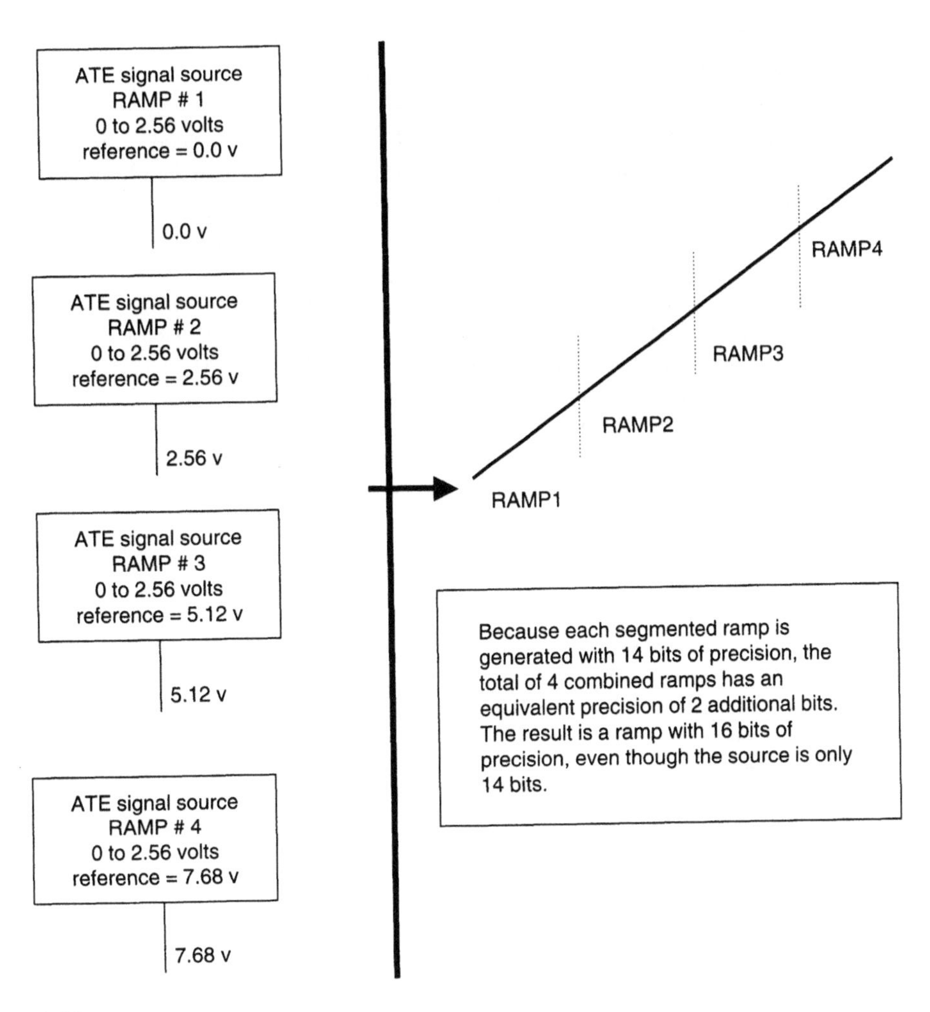

Figure 8.20

The Segmented Ramp Technique

8.7 AC Test Overview

Testing the AC performance of an ADC requires applying an analog input signal and capturing the digital output of the ADC device. Like DC testing, a typical test setup for testing the AC performance of an ADC device uses the test system signal source to generate an AC signal, such as a sine wave. This sine wave data is applied to the test system waveform generator DAC to generate a sine wave.

It is particularly important when testing AC performance that the test system synchronize the analog signal input and the strobe of the digital output with the analog clock. The test will then process the signal data with the test system's digital signal processor (DSP). As in AC testing for DAC devices, the speed of the digitizer is of greater concern than in DC testing. Test systems sometimes allow a selection of signal generators, allowing the test engineer to choose between a high-accuracy signal generator for DC tests, and a high-speed signal generator for AC tests.

The design of the test system signal generator may allow some flexibility as far as the sample rate and the number of samples per cycle. It is usually not necessary to use a power-of-two sample size for the signal generator sample set because the sample set will not be processed with a FFT. To get the best signal integrity from the ATE system signal generator, it is common practice to clock the signal generator at least 16 times the signal frequency. Any error from the signal generator can appear as error generated by the device under test.

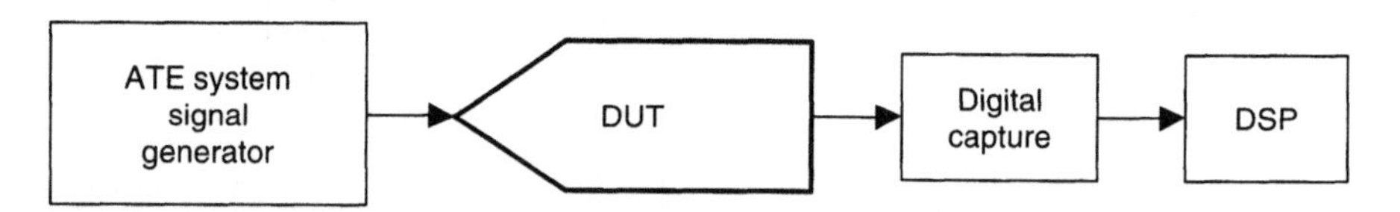

Figure 8.21

Block Diagram of ADC Test Setup for AC Parameters

8.7.1 Conversion Time

Testing conversion time consists of measuring the propagation delay from the beginning of a conversion to the expected digital output code. For some ADC architectures, such as flash converters, the conversion time can be tested in the same fashion as a propagation delay test for a digital device. Typically, the device is first conditioned by driving the analog input pin with a low voltage level, and then sequencing the device to generate a conversion. Once the output pins have been forced to a known state, the delay test consists of applying a full-scale input level and executing the device conversion sequence. The device digital output pins are strobed by the ATE system pin receivers within a specified time after the active edge of the device clock. If the device output data is correct, then the conversion executed properly within the specified conversion time.

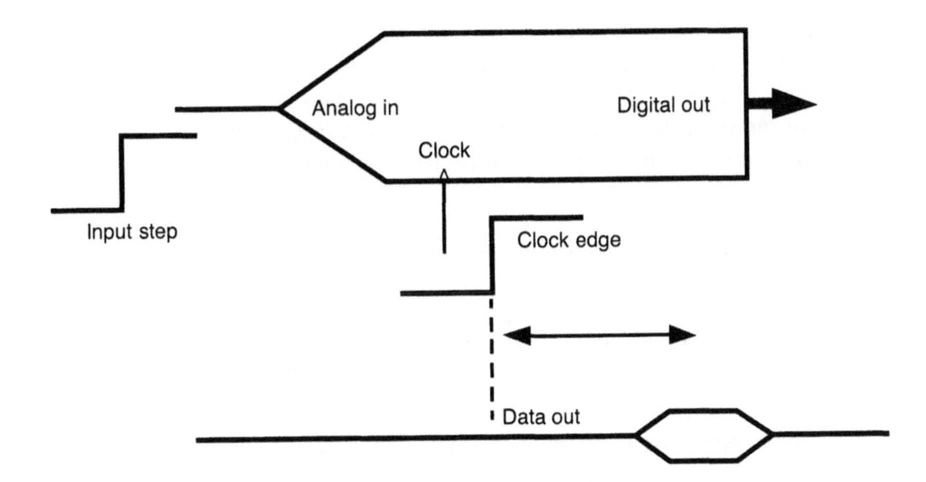

Figure 8.22

Conversion Time Test

Other converter designs, such as a successive approximation ADC, specify the conversion time in terms of the number of clock cycles required to generate the output code. In that case, conversion time is based on the specified clock count, multiplied by the clock period.

8.7.2 Harmonic Distortion Tests

The overall effects of errors across the entire range of digital codes can be expressed in terms of the ADC's ability to accurately digitize a full-scale analog signal. Testing for harmonic distortion requires that the device be driven with a analog input sine wave, synchronized with the ADC conversion rate to make sure a maximum number of unique codes are tested. The input sample set ideally would have a number of samples equal to the total number of ADC code combinations. For example, testing a 12-bit ADC would theoretically use an input pattern of 4,096 unique samples. Because there are few samples at the sine wave zero crossing, a sample set of 5 times the number of DUT code combinations is the practical minimum.

The output of the ADC must be captured via the test system digital receive circuits, and analyzed with the test system DSP. Before performing signal analysis functions, the digital data set from the ADC is converted into floating-point form. A common convention is to map the largest code value ($2n - 1$) to a positive 1.0, and the smallest code value to 0.0. A 16-bit ADC, for example, would produce a set of output codes ranging from 0000 to FFFF. Converting this data to floating point would create a data set ranging from 0.0 to 1.0, with numeric increments of 15.258e – 6 for each LSB step. The scale need not correspond to analog levels because the dynamic measurements of distortion and noise are relative measurements.

An FFT converts the captured data into frequency domain information. The frequency domain data is analyzed by first measuring the amplitude of input signal frequency, which becomes the reference point for the harmonic content ratio. The amplitude for the frequencies that are integer multiples of the signal frequency are measured and summed, and then the results are calculated as a percentage, or as a dB ratio.

$$\text{Total\_Harmonic\_Distortion} = \frac{\text{Total\_Harmonic\_Energy}}{\text{Fund}} \times 100$$

$$\text{THD\_ratio} = 20 \times \log\left(\frac{\text{Fund}}{\text{Total\_Harmonic\_Energy}}\right)$$

Interpreting the FFT Results for ADC Harmonic Distortion Tests

The test program applies a 20-kHz sine wave to the input of the ADC under test. The device is clocked at 256 kHz, and generates a total of 512 samples.

$$\text{fbase} = \frac{\text{fs}}{\text{samples}} = \frac{256\text{ kHz}}{512} = 500\text{ Hz}$$

$$\text{frequency\_bin} = \frac{\text{fi}}{\text{fbase}}$$

Table 8.1

Example Frequency Domain Data

Frequency	Frequency Bin Location	Energy Value	Energy Squared
20 kHz (fundamental)	40	0.992	—
40 kHz (2nd Harmonic)	80	1.9e–3	3.61e–6
60 kHz (3rd Harmonic)	120	8.7e–3	75.7e–6
80 kHz (4th Harmonic)	160	1.2e–3	1.44e–6

1. Find the harmonic energy as the algebraic sum of the harmonic signal energy.

$$\text{Distortion} = \sqrt{2H^2 + 3H^2 + 4H^2} = \sqrt{3.61\text{ u} + 75.7\text{ u} + 1.44\text{ u}} = \sqrt{80.75e-6}$$
$$= 8.98e-3$$

2. Calculate the dB ratio of the signal to the distortion figure.

$$-20 \times \log\left(\frac{\text{Fundamental}}{\text{Distortion}}\right) = -20 \times \log\left(\frac{0.992}{8.98e-3}\right) = -20 \times \log(110.4) = -40.86$$

You can also colculate it as a percentage.

$$\text{THD} = \frac{\text{Distortion}}{\text{Fundamental}} \times 100 = \frac{8.98\text{e}-3}{0.992} \times 100 = 0.905\%$$

Signal-to-Noise Tests

Like the DAC test sequence, the same frequency domain data set that is used to determine harmonic distortion can also be processed to derive the signal-to-noise ratio (SNR). By convention, the classic signal-to-noise measurement does not include the harmonic energy, only the non-harmonic error components. Because the SNR test is a statistical measurement, a valid number of noise components must be processed in order to generate a valid result. The capture rate (fs) and sample size must be chosen to produce a statistically valid number of data points, and a suitable bandwidth corresponding to the SNR specification.

The signal-to-noise ratio (SNR) of an analog-to-digital converter is tested by driving the DUT with a sine wave, with a known frequency and amplitude. The digital output of the device is captured and analyzed in the frequency domain by use of an FFT algorithm. Energy other than the DC and signal frequencies is a digitizing error known as noise. Testing for the SNR processes the result of the FFT to remove the energy components due to the DC value, the signal energy, and the harmonic energy. The ratio of the signal amplitude to the noise level is expressed in terms of decibels (dB).

Step One: Measure the energy value of the signal frequency (fundamental).

Step Two: Sum the noise energy

$$\text{Noise\_Energy} = \sqrt{\text{s}1^2 + \text{s}2^2 + \text{s}3^2 + \cdots + \text{s}n^2}$$

where s1, s2, s3, through sn are the frequency domain data points (frequency bins) that exclude the DC, fundamental, and harmonic signal components. A statistically valid number of noise samples, represented by the frequency bins values, is required.

Step Three: Calculate the db Ratio.

$$\text{SNR} = -20 \times \log\left(\frac{\text{Fund\_Energy}}{\text{Noise\_Energy}}\right)$$

8.7.3 The ENOBS Equation

ENOBS is one of those "insider" buzz words. You can impress your friends and neighbors by casually mentioning your estimate of the latest ADC performance in terms of ENOBs. Picture yourself by the water cooler saying, "You know, Joe, I've been looking at the spec for that new XYZ converter. It looked all right at first, but then I did a little digging. Did you know that the ENOBS is only around 10 bits?" You can bet that Joe did not know that, nor does he have a clue about what in the world an e-knob might be.

ENOBS is the *effective number of bits*, and offers another way of calculating the ratio of the signal energy to the noise energy. An 8-bit analog-to-digital converter (DAC), for example, has 256 unique digital codes. The effective resolution is 1 part in 256, or 1 LSB. Calculating the dB ratio for 1 part in 256,

$$dB = -20 \times \log\left(\frac{1}{256}\right)$$

The resolution error for an 8-bit ADC would therefore be –48.16 dB. Describing the noise ratio as an ENOB rearranges the equation to start with the dB ratio, and calculates the equivalent number of bits, normalized to a signal frequency at fs/2.

$$ENOB = \frac{SNR - 1.76\text{ dB}}{6.02}$$

Testing for signal-to-noise ratio as the effective number of bits simply calculates the value from the SNR dB measurement. When using effective number of bits as the specification, a 10-bit DAC device with a noise measurement of 48 dB would be described as "Effective Number of Bits = 8."

Table 8.2

Equivalent Number of Bits

Number of Bits	Number of Codes	Theoretical SNR (# bits × 6.02)
10	1024	–60.2 dB
12	4096	–72.2 dB
14	16384	–84.28 dB
16	65536	–96.32 dB

8.7.4 Spurious Free Dynamic Range Tests

A conventional signal-to-noise test treats the noise component amplitudes as though they were a Gaussian distribution. That works well, most of the time. However, the method of deriving the noise figure as a square root sum-of-squares can sometimes hide a significant flaw. If the frequency domain data shows a big fat noise spike in the midst of an otherwise reasonable distribution, the error will be masked somewhat by the noise calculation.

Another way of testing for noise is to look for the peak spurious component, which is the largest spectral component excluding the input signal and DC. The value is expressed in dB relative to the magnitude value of a full-scale input signal. Specification of the Spurious Free Dynamic Range (SFDR) across a given frequency is perhaps the best single indicator of the device performance. Testing for SFDR requires the device be driven with a single tone with a full-scale amplitude. The maximum frequency component excluding the signal and DC component is measured, and the dB ratio is calculated relative to the fundamental signal strength.

$$\text{SFDR} = -20 \times \log\left(\frac{\text{Signal}}{\text{Largest\_Spurious\_Signal}}\right)$$

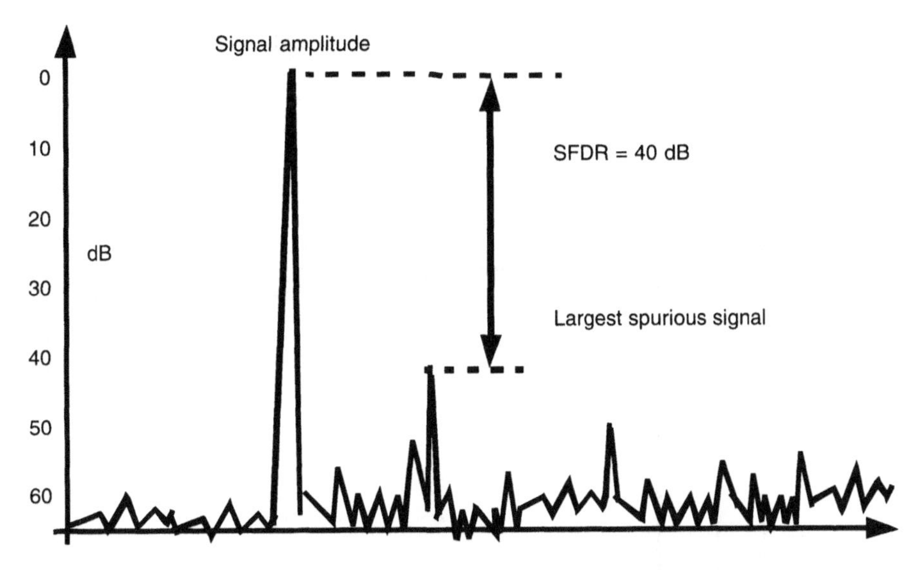

Figure 8.23

Energy Distribution for Harmonic Distortion

8.7.5 Full-Power Bandwidth Tests

ADCs exhibit some amount of frequency roll-off at higher frequencies, as a function of the input sample and hold amplifier (SHA) and the device circuit capacitance. A full-power bandwidth test verifies that the roll-off meets or exceeds the specified limit. The full-power bandwidth of an ADC is the input frequency at which the amplitude of the reconstructed fundamental is reduced by 3 dB for a full-scale input. This test is often used for devices intended for undersampling applications, where the signal is actually higher than the device clock rate.

The test uses a low-frequency signal to generate the reference amplitude. A high-frequency signal, typically much higher than the device clock rate, is applied and sampled by the DUT ADC. By measuring the amplitude of the resulting alias, the test can determine the effective roll-off introduced by the device at the high-frequency test signal.

Example

1. The test program applies a 100-kHz sine wave reference signal at full-scale amplitude to the DUT, which is clocked at 1.0 MHz. The output data set is captured and processed with an FFT. The reference signal amplitude in the frequency domain has a value of 0.997.
2. The test program applies a signal at 3.2-MHz full-scale sine wave to the DUT, which is clocked at 1.0 MHz. The 3.2-MHz alias appears at the frequency bin for 200 kHz, with an amplitude of 0.812.
3. The full-scale bandwidth attenuation at 3.2 MHz is calculated as

$$-20\times\log\left(\frac{\text{Reference}}{\text{Alias}}\right)=-20\times\log\left(\frac{0.997}{0.812}\right)=-1.78\,\text{dB}$$

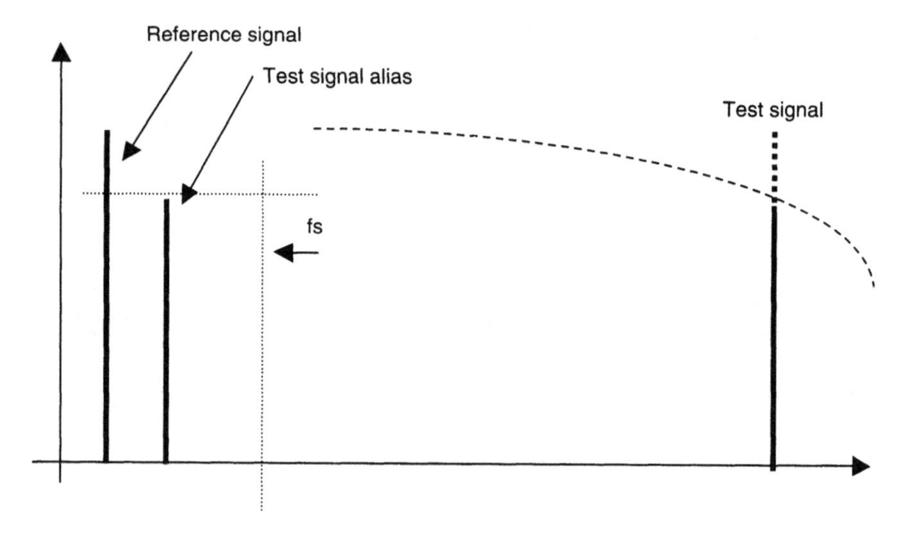

Figure 8.24

ENOBS Table

8.7.6 Aperture Delay and Aperture Jitter

Aperture delay time is the amount of time from the active edge of the device sample clock until the device actually takes a sample. Unequal delay times internal to the device cause an effective skew between the analog signal and the start of the conversion process. In practice, the aperture delay is tested by applying a series of digital pulses to the input of the ADC device under test. The device sample clock is driven with a digital edge that is swept across the aperture range. The aperture delay time is determined by monitoring the device output response. The process is similar to testing setup and hold time for a digital device.

Aperture jitter is the sample-to-sample variation in the effective point in time when the sample actually occurs. Testing for aperture jitter is performed by repeated measurements of the aperture time. A number of measurements are taken to derive the standard deviation, and the amount of aperture jitter is calculated by statistical methods. Aperture jitter can be of interest because it is directly correlated to the signal-to-noise ratio and the repeatability of high-speed digitizing applications. The effective frequency of the aperture jitter error contributes to the overall noise factor as follows:

$$\text{Induced\_SNR} = 20 \times \log\left(\frac{\text{Signal\_Period}}{\text{RMS\_Tjitter}}\right)$$

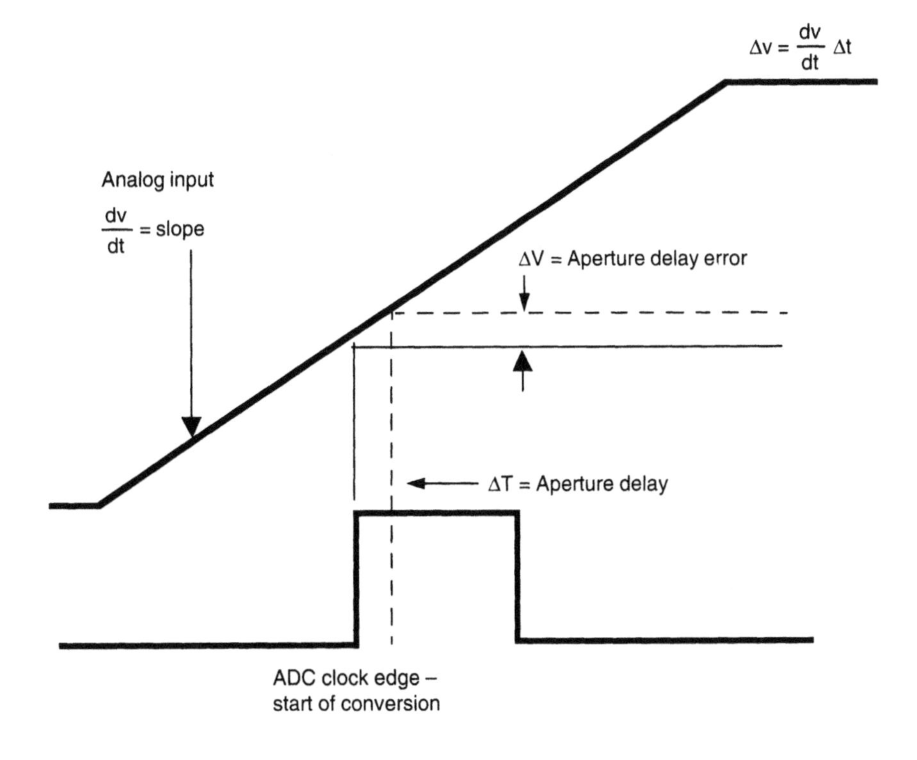

Figure 8.25

Full-Power Bandwidth Tests

8.7.7 Aperture Delay Measurement

Aperture delay can be evaluated by measuring how long the input signal must be at a valid logic level before the active clock edge. During the test, the device is programmed so the output will change state if a valid input logic state is detected. The vectors are programmed to drive the data pin with the opposite data—which forces an "all-zero" output—before and after the test data. There are two timing sets used for measuring aperture delay. The first timing set describes a default set of edge placement values, which are well within the device performance range. The second timing set uses adjustable values, which are programmed in the successive approximation process.

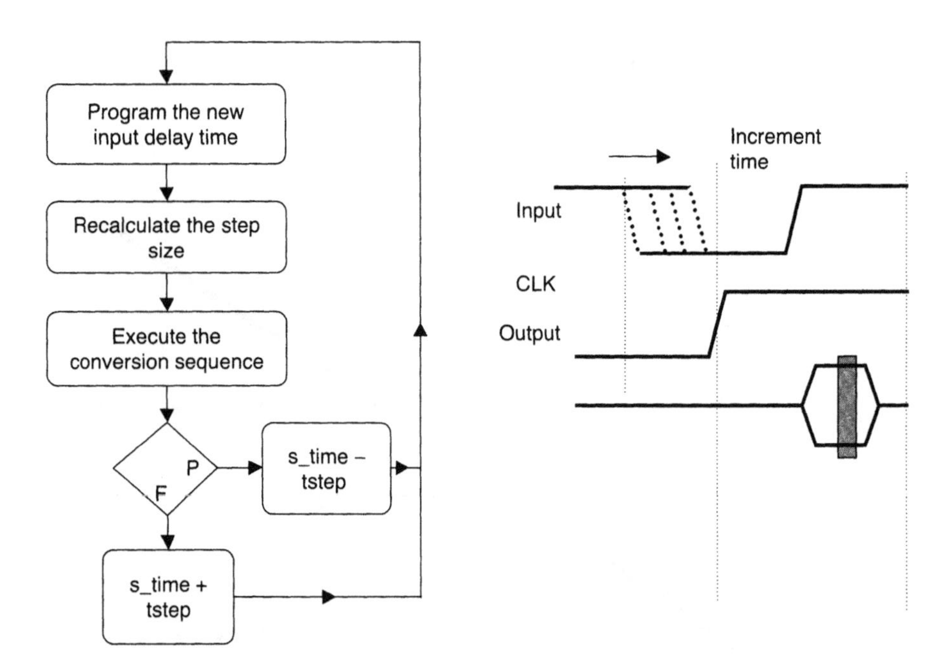

Figure 8.26

Aperture Delay

Chapter Review Questions

1. When testing analog-to-digital converters, what is meant by a "code boundary"?

2. You apply 5.0 volts to the input of an ADC device, and measure the output code at 0100. Then you apply 5.2 volts and measure the output code at 0101. Is it correct to say that the LSB increment value is 0.2 volts? (Yes or No) Why?

3. You apply a 20-kHz sine wave to the input of the ADC under test.

   The ADC under test is clocked at 256 kHz, and generates a total of 512 samples. The device output is captured and analyzed with a FFT.

   What is the frequency of the third harmonic? __________

4. The output of the FFT from question 3 can be treated as a floating-point array with the set of array elements ranging from element [0] to (fs/2)/fbase. What is the *location* of the third harmonic element in the FFT array?

CHAPTER 9

TEST CIRCUIT DESIGN CONSIDERATIONS

You can observe a lot just by watching.

—Yogi Berra

9.1 Printed Circuit Board Physics

Everybody can draw a schematic, and if you've had enough practice, it might even make sense. But getting a circuit to actually work, that's the hard part. So, what is the problem? The problem is that a physical circuit is much more complex than a theoretical circuit. The success of a mixed signal application depends on the proper design of the application software and test circuit hardware. Design of the mixed signal hardware must take into consideration requirements beyond connecting the device to the tester resources.

As an example, a 16-bit ADC with a 10-volt full-scale input has a 1 LSB value of 153 μV. To achieve an error margin of 1/2 of an LSB, the load board circuitry can generate no more than 76 μV. That's not much, bubba.

Load board effects that must be considered include

Resistance
Capacitance
Inductance
Grounding and signal routing
Power distribution
Prober and handler interface

The problem is that a physical circuit is more complex than a theoretical circuit. This additional complexity of a physical circuit is expressed by the following laws of Mixed Signal Test applications: Ohm, Kirchoff, Faraday, Lenz, and Murphy.

Ohm's Law:

The voltage drop across a resistor is equal to the current flow multiplied by the impedance, $E = I \times R$. Look, I know that you already know Ohm's law. But it bears repeating here because a little tiny amount of resistance, say, in a wire or in a socket interconnect, can introduce some serious problems.

Kirchoff's Law:

The current flowing into any point is equal to the current flowing out of the same point. The sum of voltage drops in a circuit is equal to the sum of the voltage rise. Which is to say, you don't get something for nothing.

Faraday's Law:

Two adjacent conductors separated by a dielectric are a capacitor. This is true even if two adjacent conductors are two traces on the circuit board and the dielectric is the circuit board fiberglass.

Lenz's Law:

Current flowing through a conductor creates a magnetic field.
A magnetic field that passes through a conductor creates current flow. Inductive coupling between adjacent components on a circuit board can create some interesting effects, such as noise and oscillation.
The most important law pertaining to Mixed Signal Test Applications, however, is Murphy's Law.

Murphy's Law:

In any set of circumstances, the worst thing that can happen, will.
Any effect you think can be disregarded, can't.
Nature always sides with the hidden flaw.

Example

Your 17-inch monitor has a picture tube that costs $300. The high-voltage supply of the monitor is protected by a $0.30 fuse. Applying Murphy's law, we can predict that the picture tube will explode in order to protect the fuse.

9.1.1 Trace Resistance

A resistor is not only found in those black, gray, brown, or multi-colored little cylinders with wire ends. Wires, PC traces, and socket connectors are also resistors. A common choice for PC trace width for digital device load boards is 0.25 mm. The resistance of standard PCB copper is 0.45 milli-ohms per square millimeter, so a "standard" digital PCB trace has a resistance of 18 milli-ohms per centimeter.

What would be the effect of a 5-cm length trace (standard 0.25-mm width) between the test system's analog source, and 16-bit A/D with an input impedance of 5 K?

- 5 cm × 18 millohms per centimeter = 0.09 ohms trace resistance
- 0.09 ohms/5000 ohms input impedance = 0.0018% gain error

The trace resistance acts like a divider resistor in series with the device input impedance, creating a voltage drop equal to 1 part in 55,000 (1:55 k), or 0.018%. In comparison, an analog-level increment of 1 LSB for a 16-bit A/D converter is 0.0015% of full scale. There is more error introduced by the trace resistance than is generated by the device.

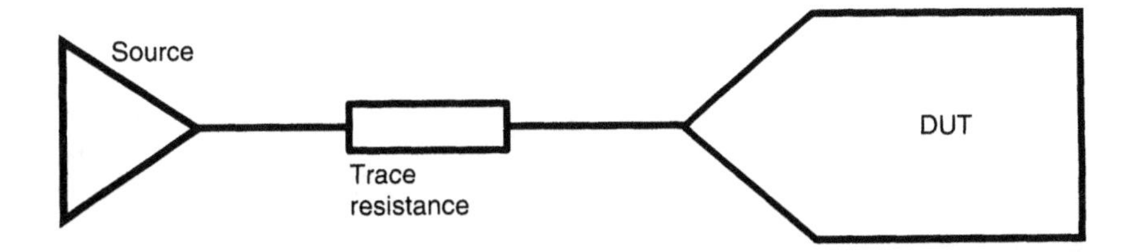

Figure 9.1

PCB Trace Resistance

Solution:

Every doubling of the trace width reduces the resistance by half. Use wide traces whenever possible. To correct for the voltage drop, use Kelvin connections from the tester resources wherever possible. In applications where the trace resistance is critical, measure the resistance with a milli-ohmmeter, and compensate for the error in software.

9.1.2 Force and Sense Connections

The ATE system signal source hardware may include force and sense connections, which allow the source to correct for any voltage drop due to trace resistance. The output of the source circuit behaves like an op-amp configured as a *force compliance servo loop*. The op-amp will drive its output to be whatever is required to balance the two input levels. To force compliance with a voltage drop in the force line, the op-amp output will generate a voltage offset to correct for the difference between the reference and the sense line.

Usually, the best results are achieved by connecting the force and sense lines together as close to the device pin as possible. This allows the Kelvin circuit to compensate for any voltage drop caused by trace resistance on the load board.

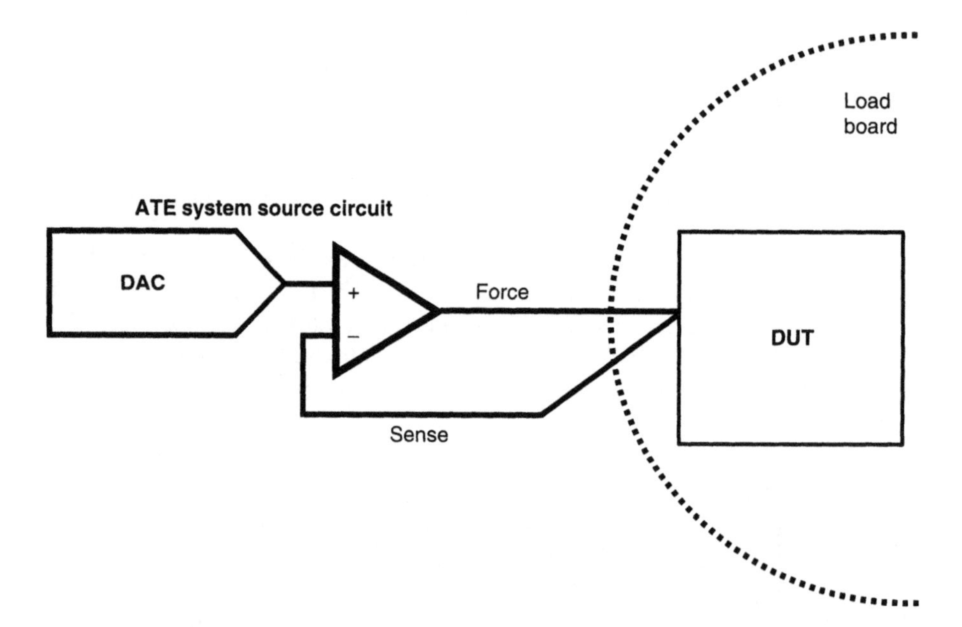

Figure 9.2

Force and Sense Connections

9.1.3 Skin Effect

Inductance at high frequencies has the effect of causing current to flow only on the surface of the conductor, which increases the effective resistance. The amount of resistance increases by the square root of the frequency. At the typical printed circuit trace thickness of 0.038 mm, the skin resistance of a 0.25 trace per centimeter can be calculated as

$$\text{SkinR} = 2.6 \times 10^{-7} \times \sqrt{\text{freq}}$$

Example

9.0 MHz 8.2 milli-ohms per centimeter
100 MHz 26.0 milli-ohms per centimeter
1000 MHz 82.2 milli-ohms per centimeter

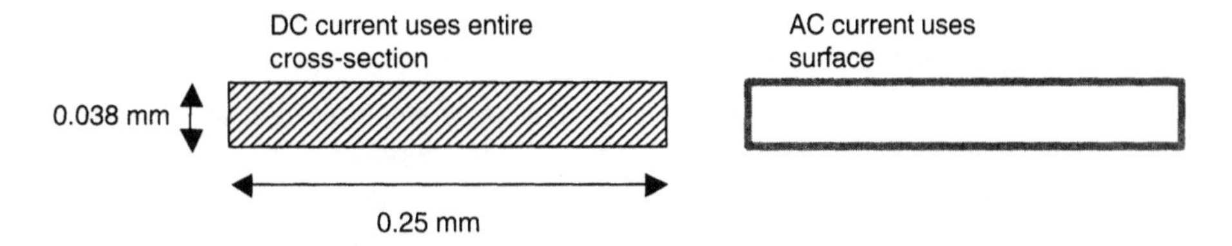

Figure 9.3

DC and AC Current Trace Current

High-frequency current flows only in surface layers, and wide traces provide a large surface area. Particularly for high-current transients such as those found in a ground return connection, a large surface area reduces the skin effect impedance.

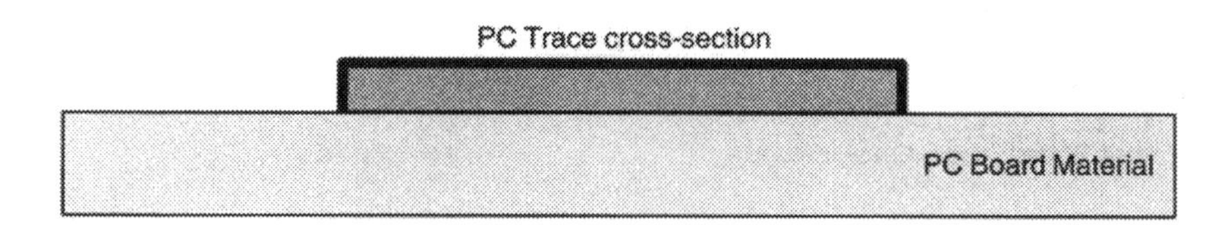

Figure 9.4

PC Trace and PCB Cross-section View

9.1.4 Circuit Board Inductance and Capacitance

All conductors are inductive. The inductance of a standard 0.25-mm PCB trace is about 10 nH per centimeter. The formula for inductive reactance is

$$X_L = 2\Pi f L$$

Therefore, at 10 MHz, a PC board trace length of one centimeter has a reactance of 0.62 ohms. For a 50-ohm system this can create a greater than 1% error. Inductance is directly related to the length and surface area of the conductor. Short wide traces have the lowest inductance. Long narrow traces have high inductance.

An inductor in series or in parallel with a capacitor forms a resonant, or "tuned," circuit. A resonant circuit can cause oscillations. If the circuit is behaving erratically, but the problem seems to disappear when you apply a scope probe to the circuit node, chances are you have a tuned circuit. What happens is the capacitance of the scope probe retunes the resonance to a less troublesome frequency. Spurious oscillations of this sort can usually be corrected by adding surface-layer ground plane, or re-orienting the trace layout in the specific circuit section.

Circuit Board Capacitance

Capacitance can exist between two adjacent conductors that are separated by an insulator. If there is a change in voltage on one conductor, there will be a change in charge on the other. General-purpose PCB material has a capacitance of about 3.0 pF per square centimeter for conductors on opposite sides of the board. A grounded conductor between the two "plates" of a capacitor functions as a Faraday shield. The shield must be connected to block the coupled noise and allow the noise current to return to ground without flowing through a sensitive section of the circuit.

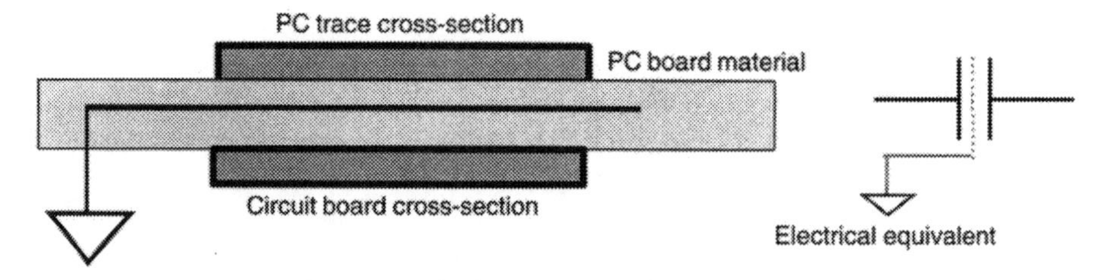

Figure 9.5

Circuit Board Trace Capacitance

9.2 Resistor Physics

Discrete component resistors include some amount of capacitance and inductance, as well as an effective noise generator based on the junction of two dissimilar materials.

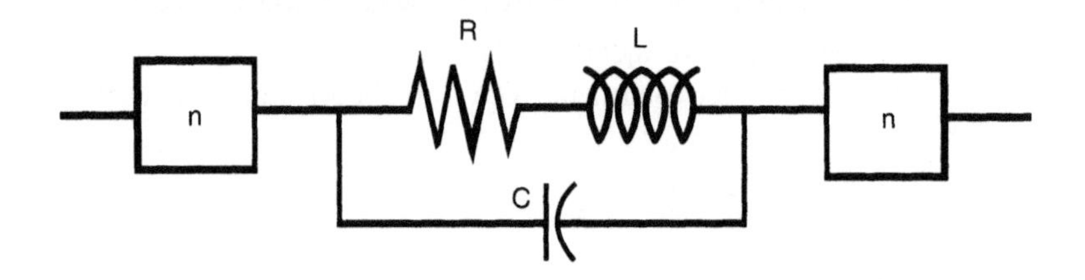

Figure 9.6

Resistor Physics

- The capacitance (C) of a resistor is typically around 5 pF.
- The inductance (L) is typically around 1 or 2 uH.
- The thermoelectric effect (n) of the ni-chrome to copper/nickel junction of a wirewound resistor is 42 μV/C.

Thermal noise, also known as Johnson noise, is found by

$$\mathrm{Vnoise} = \sqrt{4kTBR}$$

where

T = absolute temperature (in degrees Kelvin),
B is bandwidth in Hz,
R is the resistance in ohms, and
k is Boltzmanns' constant (1.38e − 23 J/K).

Reducing the resistance, the bandwidth, or the temperature will reduce thermal noise.

9.3 Capacitor Physics

A capacitor as it behaves in a mixed signal application is actually a complex parallel network of capacitance, resistance, and inductance.

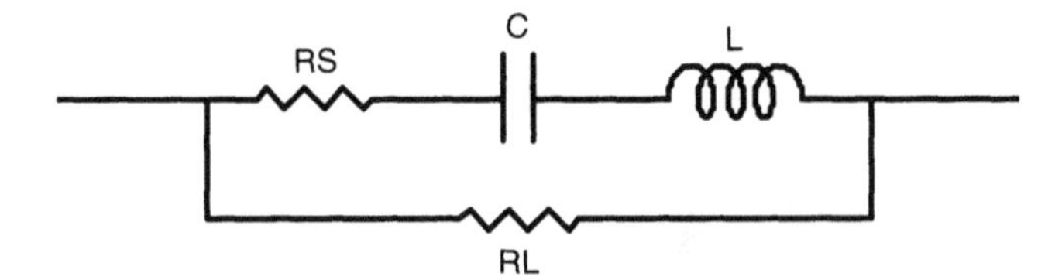

Figure 9.7

Capacitor Physics

The leakage resistance (RL) illustrates the leakage current path. Electrolytic capacitors have a current leakage of about 20 nA per uF. Tantalum capacitors have a much better leakage current of around 5 na per uF. Besides, if you wire an electrolytic capacitor in backwards, when it explodes, it makes a big mess and bad smell. You end up with confetti and oil all over, along with the smell of burnt squash. On the other hand, if you wire a tantalum capacitor backwards, it will emit a column of flame and smoke, which is really quite dramatic, but the smell tends to dissipate much more quickly.

The series resistance (RS) illustrates the power dissipation that occurs in capacitors. This can be a concern with RF applications, and power supply decoupling caps that have a high ripple current.

The series inductance (L) of a capacitor is created by the physical structure. In general, electrolytic, paper, and plastic film capacitors may tend to behave more like inductors at frequencies above 1 MHz. Monolithic ceramic caps have very low series inductance, and are usually a good choice for high-frequency applications.

9.4 Circuit Board Insulators and Guard Rings

The insulator resistance on a F4 fiberglass PCB varies from one section of the board to another. The resistance laterally (such as between two plated feed-throughs) is usually much lower than the resistance between two surface traces. A clean PC board will usually have an insulation resistance of between 10 and 100 Mega Ohms per cm. Besides worrying about ground currents and supply currents, sometimes we have to worry about the current that is sloshing around on the surface of the circuit board fiberglass. In applications that involve very high impedance and very low currents, the sensitive circuit section can be isolated from the effects of board leakage by a guard ring. A guard ring is a circular trace around the sensitive circuit point, which is driven to the same voltage level. A guard ring on both sides of the board will create the condition of no voltage drop across the protected circuit area, which in turn creates the condition of no current flow from leakage resistance.

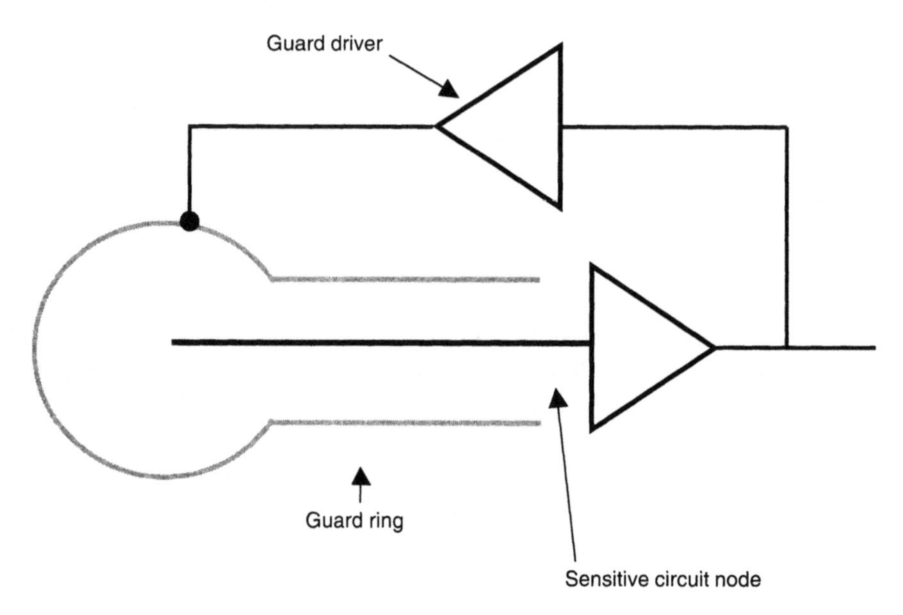

Figure 9.8
Guard Rings

9.5 Test Circuit Ground

In a physical circuit, GROUND is not a schematic symbol. GROUND is a *current return path*. Other than the lie about how being an engineering major would make it easy to meet attractive people of the opposite sex, this is probably the biggest lie we were told in engineering school.

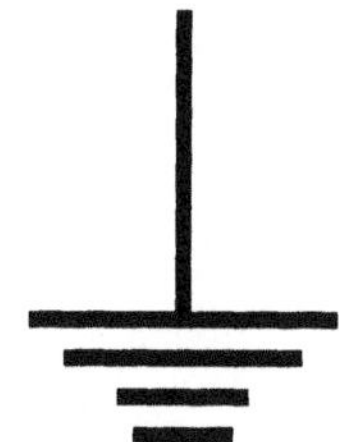

Figure 9.9

Phony Baloney Ground Symbol

The ground symbol has little or nothing to do with the actual flow of ground current. A friend of mine likes to say, "Ground is like the ocean." Really, he actually says that. If there is anything like an ocean in an accurate analog of ground, it would be "Ground is like a two-lane dirt road to the ocean on a summer afternoon." As in, lots of traffic, backups, and collisions.

Kirchchoff's law tells us that current has to come from somewhere, and has to go somewhere. Every variation in current from the power supply and device I/O pins is reflected in the device ground as a current return path. Because a mixed signal test circuit ground is implemented with wire and circuit board copper, all of the physical effects of an actual conductor also apply to ground—particularly resistance and inductance. Ground might theoretically be ZERO volts, but it is not ZERO current.

Transient ground current can be much greater than the specified average IDD current. A 16-bit ADC driving 100 pF on each output pin with a 5-volt 5-ns swing can generate 1.6 amps of transient current.

$$\frac{\Delta \text{ Voltage}}{\Delta \text{ Time}} \times \text{C} = \text{I}$$

$$\frac{5.0 \text{ v}}{5 \text{ ns}} \times 100 \text{ pF} = 100 \text{ mA} \times 16 = 1.6 \text{ A}$$

All of the transient current will be present in the ground return path. Any impedance in the ground return path will create a voltage drop that is sometimes called "ground bounce."

9.5.1 Ground Loops and Shared Ground Current

A ground loop problem can occur when a circuit has more than one return to ground, and that ground return path is shared with another circuit. The resistors illustrate the impedance of the ground connection path.

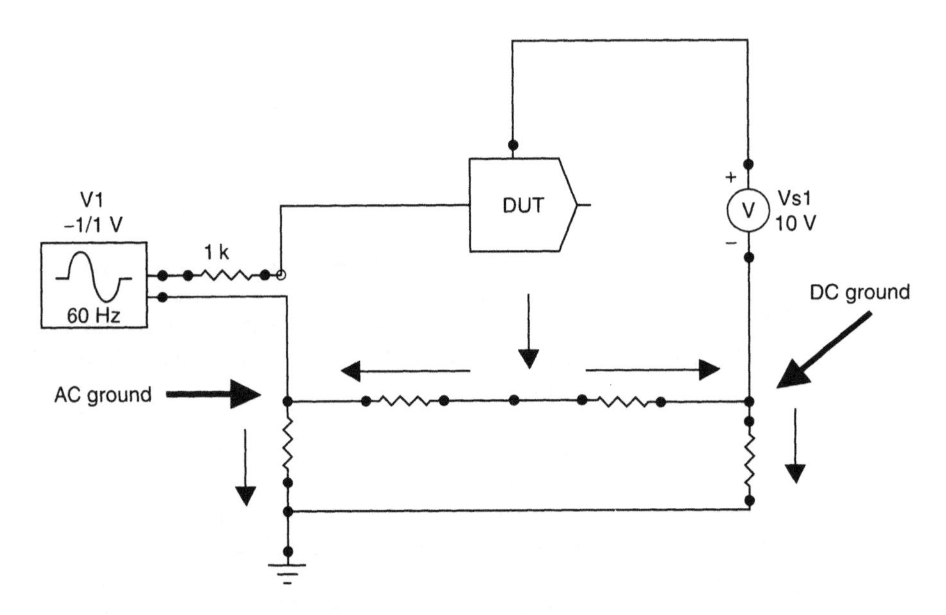

Figure 9.10

Ground Loop

In Fig. 9.10, both the AC signal generator and the 10-volt DC source have a path to ground with some finite impedance, resulting in a voltage drop. The DUT ground is connected, via a finite impedance, to both the "AC Ground" and the "DC Ground." Both the AC and DC ground currents create an offset voltage at the DUT ground point. There are different ways to represent a ground loop. In essence, a ground loop occurs when an external circuit shares the same ground return path as the signal circuit. A serial (or "daisy-chained") ground can cause the same types of problems as a ground loop.

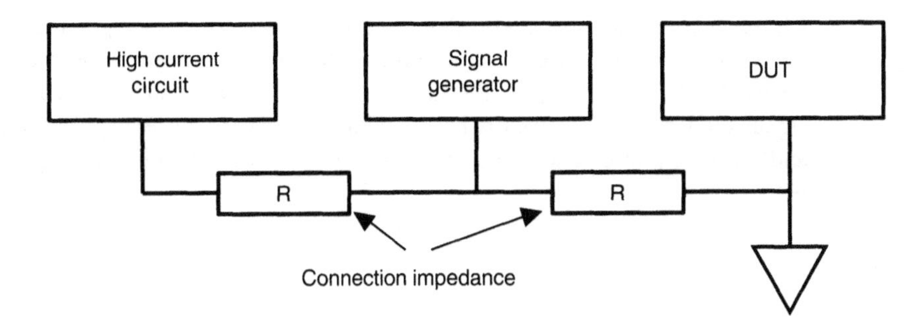

Figure 9.11

Shared Ground Return

9.5.2 Star Ground

One approach for avoiding shared ground paths is to design the ground scheme so that each circuit section has a separate path to ground. This produces a single point in the circuit that serves as the ground, to which all circuit voltages are referenced. That single point is known as the "star" ground.

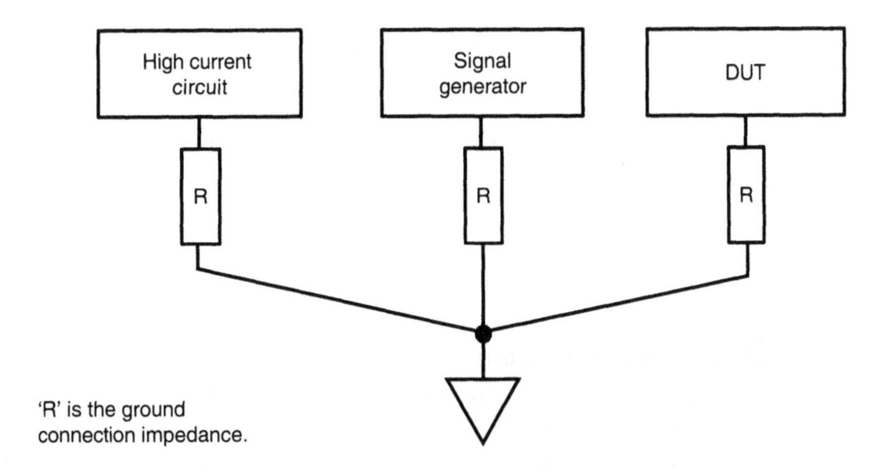

Figure 9.12

Star Ground

The star ground is an important concept that is often combined with other techniques to achieve an optimal grounding design. One of the problems that must be addressed is that the power supplies have their own ground paths. The supply current that flows through the existing ground paths may be of a high value, or just noisy. Kirchoff's Law applies to AC currents as well as DC, and it certainly applies to ground current. Any transient current flow on the supply, such as may be caused by digital circuit activity, is mirrored in the ground current.

Digital circuitry is noisy, and is designed to tolerate noise. Analog circuitry is quiet, and cannot tolerate noise. Circuit noise in the form of current transients carried by power supplies and the ground returns can create errors in analog circuits. In mixed signal applications, it is usually best to dedicate separate power supplies to the digital and the analog circuit sections. Each supply should have a dedicated ground return path.

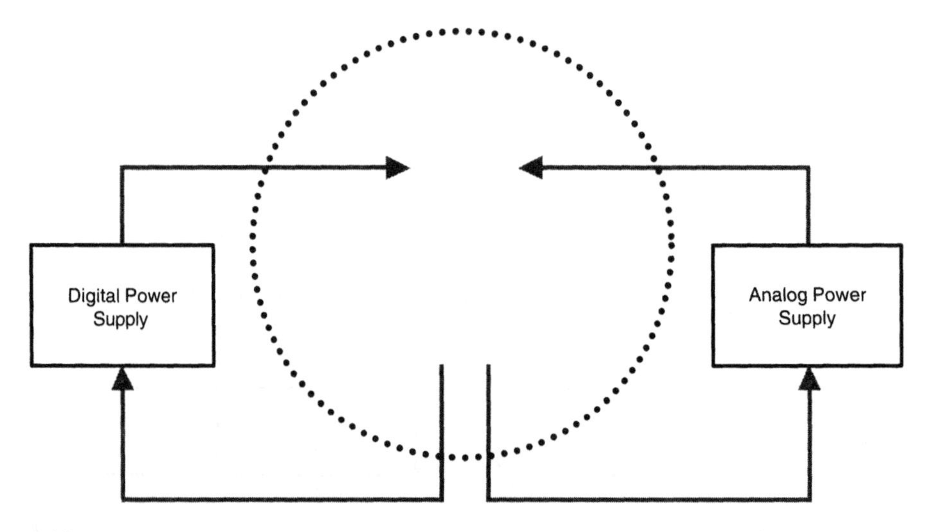

Figure 9.13

Dedicated Ground Returns

9.5.3 The Ground Plane

Because the lowest impedance path is achieved with the widest possible trace, a trace that covers a large portion of the circuit board is called a "plane." The ground plane is simply a very wide trace that allows a low impedance current return path. Analog and digital grounds must be connected at one point, which serves as the star ground reference.

The test circuit ground planes serve several purposes. The primary function of a ground plane is to provide a low impedance current return path. The ground plane can also function as a Faraday shield between signal traces, either between circuit board layers or between adjacent traces on the same board.

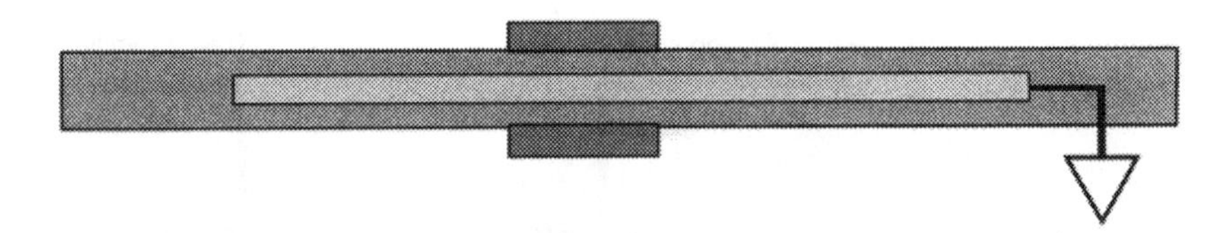

Figure 9.14

Center Layer Ground Plane

The inductive field generated around components that are mounted on the surface of the circuit board can couple into adjacent components. A ground plane on the surface of the circuit board decouples the inductive fields.

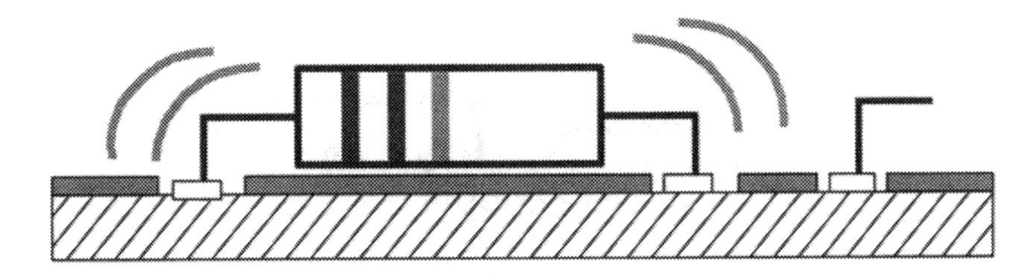

Figure 9.15

Surface Layer Ground Plane

9.5.4 Split Grounds

Many mixed signal devices have separate supplies and separate ground pins for the analog and digital grounds. The data sheets on these devices usually advise to connect these pins together at the device package. It is usually best to heed this advice. The problem the device manufacturer is trying to address concerns the limitation of the device package technology. The voltage drop in the bond wires is too large to allow the connection to be made internally.

The labels "analog ground" and "digital ground" on the device pins refer to the sections of the device to which the pins are connected. The labels of "analog ground" and "digital ground" are not intended to communicate the ATE system ground preference. (You may find it surprising to learn that design engineers do not stay awake late at night trying to figure out how to make life easy for the test engineer.) Optimally, it is usually best to conform to the DUT manufacturer's advice and tie the device digital and analog grounds together at the DUT, making that point the star ground. If that is not feasible, all of the DUT grounds should be tied to the analog ground.

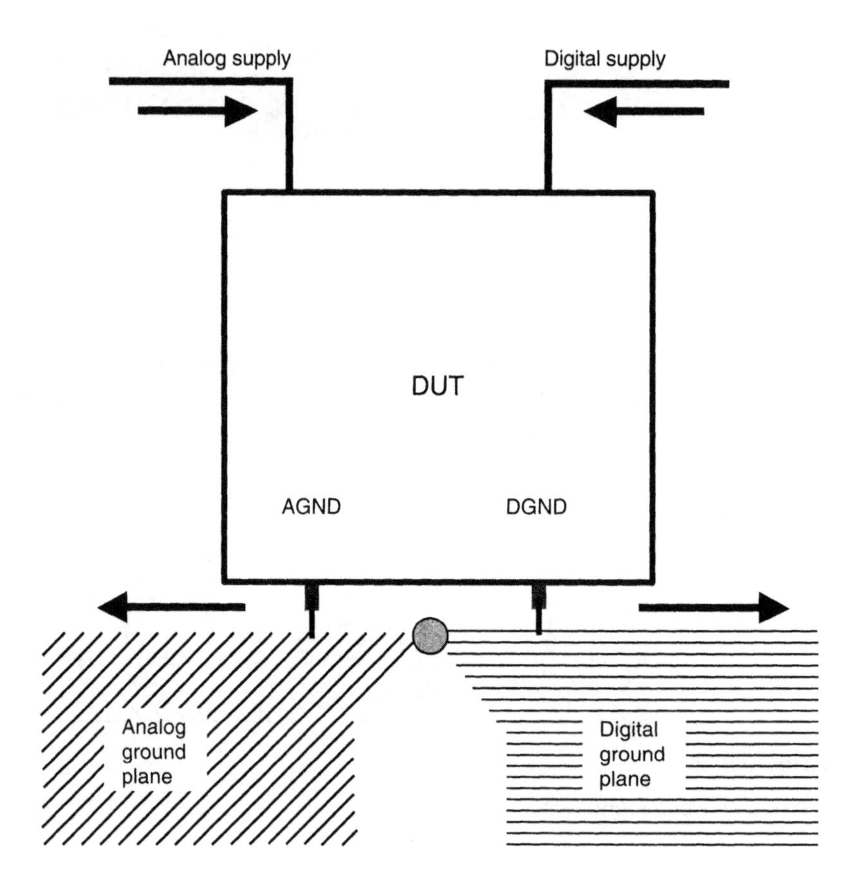

Figure 9.16
Split Grounds

9.6 Power Distribution

The power connections to the device under test must also provide a low impedance path for the supply current. The same principles that apply to ground current paths also apply to power supply current paths. As with the ground connections, a very wide trace area or *plane* provides the lowest resistance and inductance. Circuit board planes that supply current are known as power planes. The transient power required by the DUT is much higher than the average DC current. Even though the VOH/IOH and VOL/IOL specifications may indicate a higher level of equivalent impedance, the transient impedance is much lower, and transient current for digital output pins is much higher. Reviewing the previous example, a 16-bit ADC driving

100 pF on each output pin with a 5-volt 5-ns swing can generate 1.6 amps of transient current.

$$\frac{\Delta \text{ Voltage}}{\Delta \text{ Time}} \times C = I$$

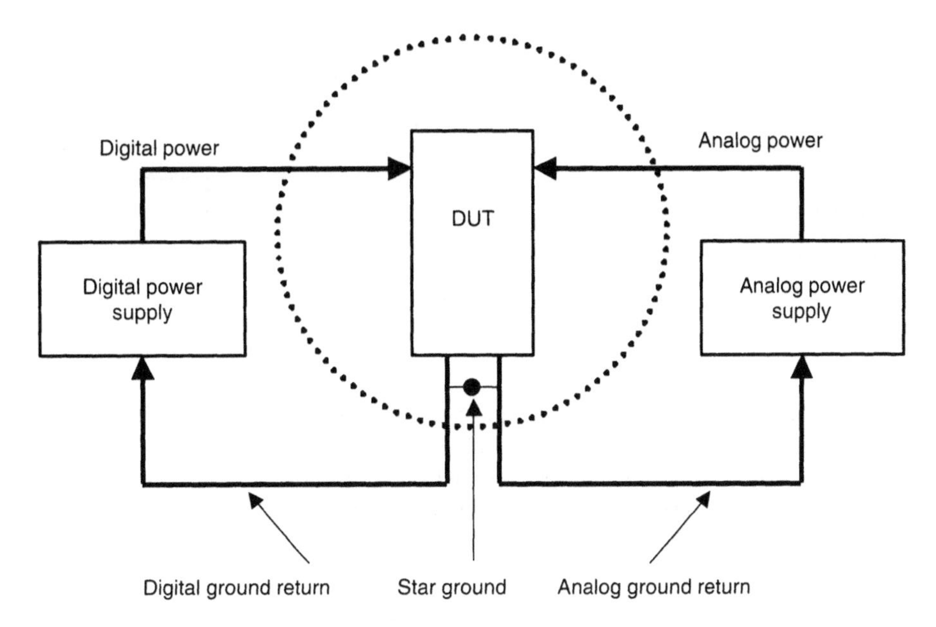

Figure 9.17

Power Distribution

9.6.1 Power Supply Decoupling

Most ATE systems are designed with the power supplies located in the mainframe connected to the test head with a long cable. In order to supply large amounts of current in a short amount of time, the power supply must have a local current reservoir as close to the device as possible—the bypass capacitors.

Mixed signal applications typically use a combination of bypass caps to address the low-frequency and high-frequency current demands. Capacitors with larger values tend to have larger values of associated inductance, so smaller caps in parallel provide a compound solution. A 10-uF tantalum cap in parallel with a 1.0-uF and 0.1-uF monolithic cap is a common choice.

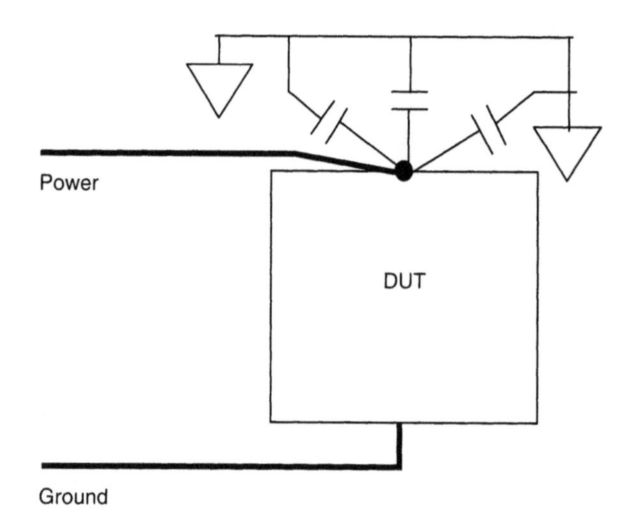

Figure 9.18

Power Supply Decoupling

The 0.1-uF capacitor has the lowest inductance, and responds immediately to the current demand. The 1.0-uF capacitor takes longer to respond, but can supply current for a longer period of time. The 10.0-uF capacitor has the highest inductance and the slowest response time, but can supply current for the duration of the transient current demand.

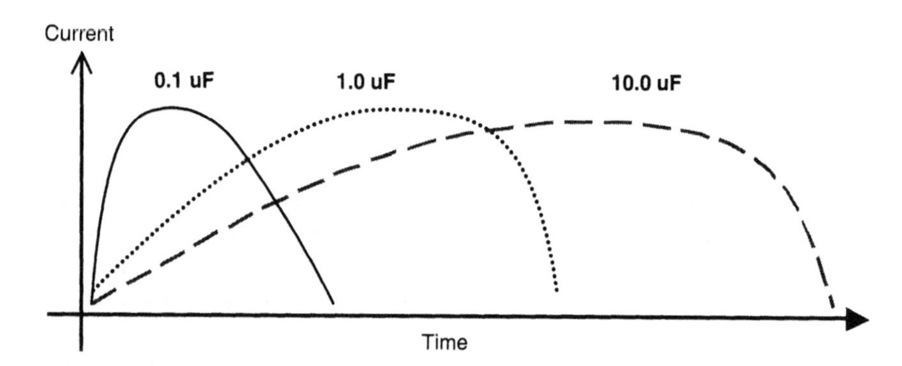

Figure 9.19

Overlapping Current Supply

Test Circuit Summary

- Think about where the currents go.
- Conductors have resistance, inductance, and capacitance. So do resistors, capacitors, and inductors.
- Ground is a signal path.
- Current that flows out of the device must come from somewhere.
- Current that flows into the device must go somewhere.
- Keep PC traces short and wide.
- Use ground planes, especially under components and traces that operate at high frequency.
- Use separate supplies and separate returns for the digital and analog sections. Tie the returns together at a single point.
- Physically separate analog and digital signal traces.
- Avoid crossovers between analog and digital signals.
- Be *very careful* with high impedance points. Consider the use of guard rings.

9.7 Transmission Lines

A transmission line is an interconnect consisting of two conductors and a dielectric. Characteristics include distributed series resistance and inductance, along with distributed parallel capacitance between the two conductors. Coax cable is a common example of a transmission line, but the same effects are evident with circuit board interconnect paths.

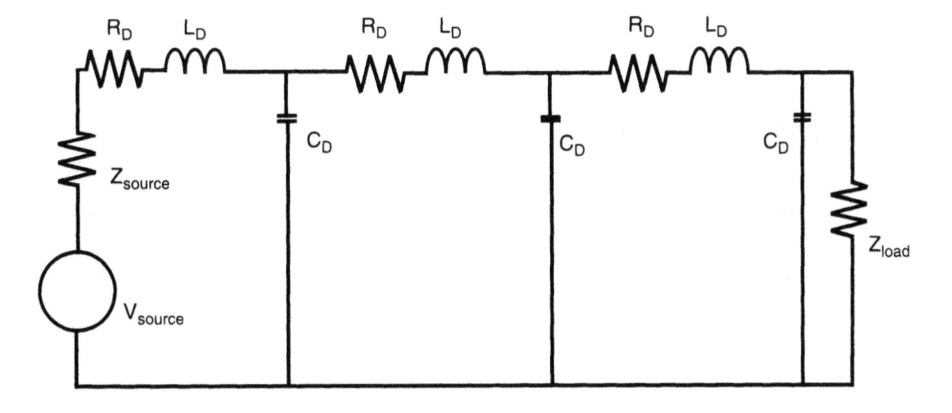

Figure 9.20

Transmission Line Circuit

Transmission line impedance mismatch can result in signal delays or ringing at interconnects, which in turn cause timing errors or incorrect data.

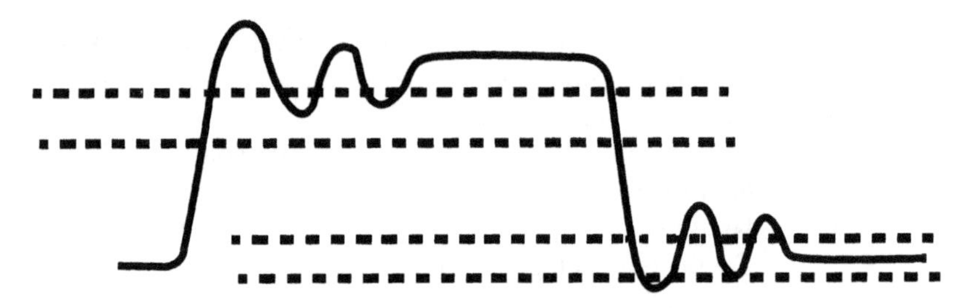

Pulse with ringing, overshoot, and undershoot caused by transmission line mismatch.

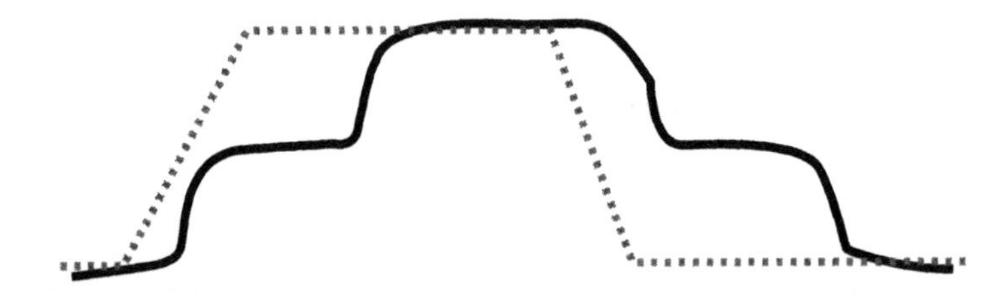

Stair-step effect and pulse delay caused by transmission line mismatch.

Figure 9.21

Transmission Line Effects

9.7.1 Transmission Line Reflections

Transmission line effects are basically Ohm's law across time. It takes a finite amount of time for a pulse to travel from one end of wire to the other. Between the time when the pulse is generated and when it gets to the end of the wire, the only impedance is the wire itself. The wire, or transmission line, impedance acts as a current load for a period equal to one round trip delay for each signal transition. The transmission line does not present any current load under DC conditions. If the load impedance does not match the transmission line impedance, the current differential causes a reflection from the load back to the source.

If the current differential is extreme, a transmission line with mismatched impedance can cause signal voltages to appear to double at the receiver termination.

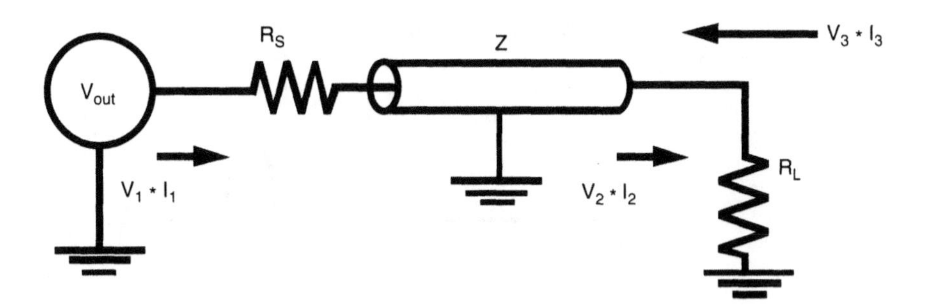

Figure 9.22

Simplified Transmission Line

1. The initial condition of voltage and current is equal to V_1 and I_1 at the signal source.
2. The condition of voltage and current at the end of the transmission line is equal to V_2 and I_2 at the signal termination.
3. If the source and termination condition of voltage and current are not the same, the transmission line reflections are equal to V_3 and I_3.

The initial condition current will be equal to: $I = \frac{V_1}{Z}$

The final condition current will be equal to: $I = \frac{V_2}{RL}$

Example

Circuit Conditions:

V_{out} = 0 to 3.0 volts transition Z = 50 ohms RL = 10 K

Transmission Line Impedance = 50 ohms Transmission line propagation delay = 5 ns

For the first 5 ns after the signal transition, the voltage source drives only the 50-ohm transmission line impedance. Driving 3 volts into 50 ohms requires 60 mA. This 3 volts at 60 mA flows down the transmission line until it encounters the termination of RL. Because the load impedance of 10 K across 3 volts absorbs only 0.3 mA, a reflected wave must be generated to satisfy Ohm's Law. The reflected current goes back across the 50-ohm transmission line. The reflected current of 59.7 mA across the 50-ohm impedance produces a voltage step ($E = I \times R$) of 2.98 volts, which is superimposed on the 3-volt signal from the source. The signal level effectively doubles because of the reflected current. The current continues to bounce back and forth between the source and the load until it is absorbed. Guess what? That's what causes ringing.

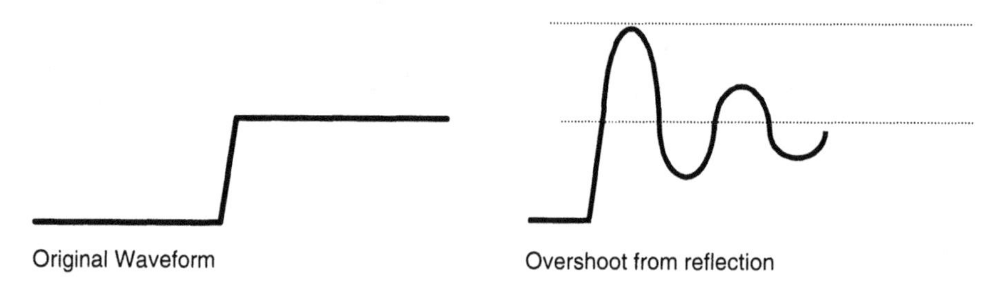

Figure 9.23

Transmission Line Overshoot and Undershoot

The properties of the transmission line are described by the impedance (Z) and the propagation delay (Tpd). Both the transmission line impedance and the propagation delay are a function of the distributed inductance (L) and capacitance (C).

$$Z = \sqrt{\frac{L}{C}} \qquad Tpd = \sqrt{LC}$$

The reflection coefficients are a function of the ratios of source impedance, transmission line impedance, and the load impedance.

Load Reflection Coefficient: $PL = \dfrac{Rload - Z}{Rload + Z}$

Source Reflection Coefficient: $PS = \dfrac{Rout - Z}{Rout + Z}$

There are several combinations of source, transmission line, and load impedance that may occur in a test circuit.

9.7.2 High Impedance Load Effects

1. A high-speed device output with corresponding high-current capability and a low value of Rout (5 ohms)

   connected to
2. A 50-ohm transmission line

   connected to
3. The input of a high-impedance buffer, with an input impedance of 1 MB ohm.

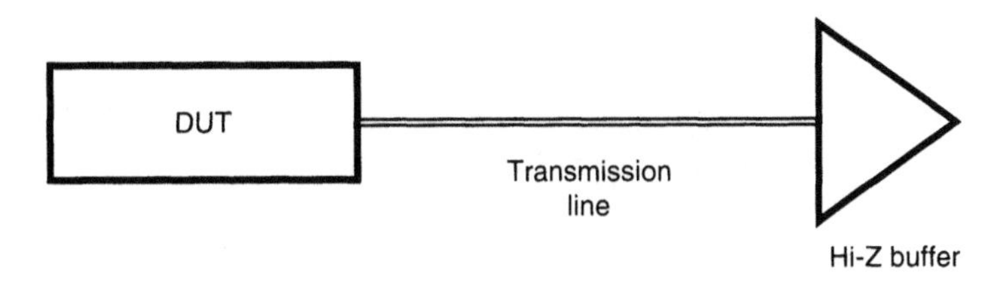

Figure 9.24

High Impedance TL Circuit

Calculate the load reflection coefficient.

$$PL = \frac{Rload - Z}{Rload + Z} = \frac{1\_Meg - 50\Omega}{1\_Meg + 50\Omega} = 0.99$$

Calculate the source reflection coefficient.

$$PS = \frac{Rout - Z}{Rout + Z} = \frac{5\Omega - 50\Omega}{5\Omega + 50\Omega} = -0.81$$

A transmission line with a positive load reflection coefficient (PL) will exhibit ringing, overshoot, and undershoot. A matched transmission line will approximate zero for both the load and the source reflection coefficients.

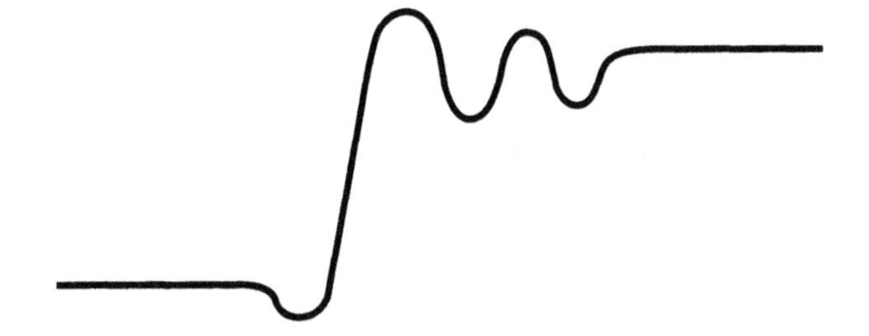

Figure 9.25

High Impedance TL Signal Results

9.7.3 Low Impedance Load Effects

In some test circuit implementations, the transmission line may be terminated with a low impedance load.

Example

1. A high-speed device output with corresponding high-current capability and a low value of Rout (5 ohms)

   connected to
2. A 50-ohm transmission line

   connected to
3. A parallel-terminated, low-impedance input (10 ohms).

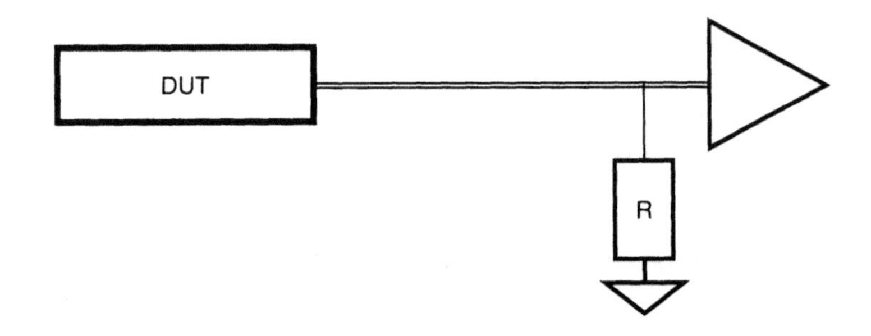

Figure 9.26

Low Impedance TL Circuit

Calculate the load reflection coefficient.

$$PL = \frac{Rload - Z}{Rload + Z} = \frac{10\Omega - 50\Omega}{10\Omega + 50\Omega} = -0.66$$

Calculate the source reflection coefficient.

$$PS = \frac{Rout - Z}{Rout + Z} = \frac{5\Omega - 50\Omega}{5\Omega + 50\Omega} = -0.81$$

A transmission line with a negative load reflection coefficient (PL) will exhibit a stair-step distortion or improper duration of the pulse period.

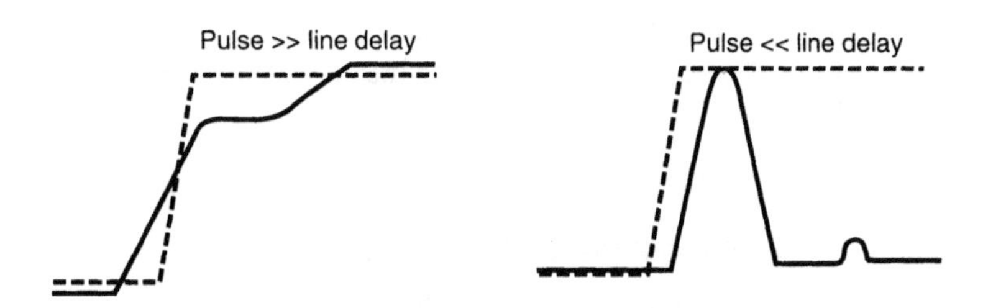

Figure 9.27

Low Impedance TL Signal Results

I ran into this phenomenon once with a video amp application, and it nearly drove me crazy. The output of the DUT was connected to a high-current buffer, which in turn drove a terminated coax cable to the receive circuit.

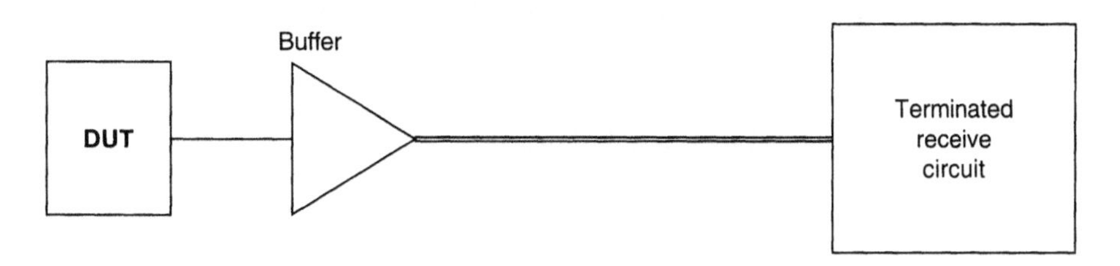

Figure 9.28

Buffered Signal to Mis-terminated Load

On the output of the DUT, the scope showed a nice, fat, reasonable-looking pulse. On the other side of the buffer, the signal had degraded into a skinny little glitch. My first instinct was to blame the hardware, so I replaced the buffer. That didn't help, so I had to get more scientific. I noticed that if the buffer output was not connected to the transmission line, then the signal coming out of the buffer looked just fine. That gave me enough clues to suspect the transmission line, and here's what I found:

Output Buffer Impedance:	1 ohm
Transmission Line Impedance:	50 Ohms
Receive Circuit Termination:	37.5 ohms (75 ohm video load, double terminated)

Well, yeah, it is obvious now, but it sure wasn't then. Let's take a look at the coefficients.

$$PL = \frac{Rload - Z}{Rload + Z} = \frac{37.5\Omega - 50\Omega}{37.5\Omega + 50\Omega} = -0.142$$

$$PS = \frac{Rout - Z}{Rout + Z} = \frac{1\Omega - 50\Omega}{1\Omega + 50\Omega} = -0.96$$

Yep, that will do it, all right. In this case, I was able to change the receive impedance from 37.5 ohms to 50 ohms, and then re-calculate the expected voltage and current levels.

9.7.4 Edge Rate and Line Length

The effects of the transmission line vary based on the edge rate of the DUT and the propagation delay of the transmission line. If the device edge rate is short compared to the transmission line delay, the effects of the impedance mismatch become apparent. If the rise and fall time of the device is long in comparison to the transmission line delay, the effects are hidden during the transition time. In general, a PC board trace will act like a transmission line if the rise and fall time of the device is less than three times more than the transmission line delay. A circuit board trace designed to function as a transmission line is known as a "strip line." The propagation delay for a strip line is approximately 2.5 ns per foot.

9.8 Transmission Line Matching

Designing a matched transmission line must take into account the source resistance, the transmission line impedance, and the load resistance. The source resistance must be less than or equal to the transmission line impedance (Z) to minimize the effect of the source coefficients. Secondly, the load resistance (RLoad) must be equal to the transmission line impedance (Z). A properly matched transmission line circuit has a load reflection coefficient (PL) of 0. There is no ringing or stair-step effect.

There are several alternatives for matching the transmission line to the source impedance and the load impedance, as follows:

- Series Termination
- Parallel Termination
- Thevenin Termination
- RC Network Termination
- Diode Termination

(*Note:* The following examples use the abbreviation "TL" for Transmission Line.)

Also known as back matching, *series termination* inserts a resistor in series with the source to effectively change the value of Rout. This method has the advantage of simplicity, and is often the termination of choice when the transmission line impedance and the load impedance are fixed. The primary disadvantage is that series resistance (RB) limits the amount of available current, which can adversely affect the signal rise and fall time.

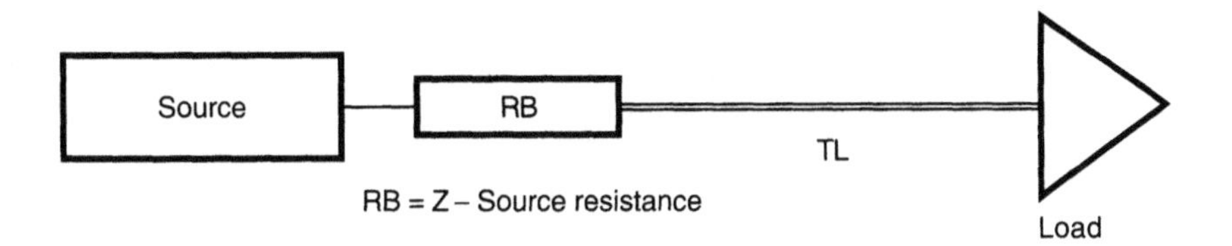

Figure 9.29

Series Termination

Parallel Termination Inserts a resistor in parallel with the RL at the end of the transmission line to change the value of RL. Parallel termination allows efficient absorption of the transient current, with no adverse effect on the signal transients. The two disadvantages are the constant DC load on the source, and asymmetrical loading. Asymmetrical loading is caused by a larger current demand for source high levels than for low levels.

Example

Source VOH = 3.0 volts
Source VOL = 0.4 volts
The source current when driving a logic high is equal to the VOH level divided by the termination resistance: 3.0 volts/50 ohms = 60 mA.
The source current when driving a logic low is equal to the VOL level divided by the termination resistance: 0.4 volts/50 ohms = 8 mA

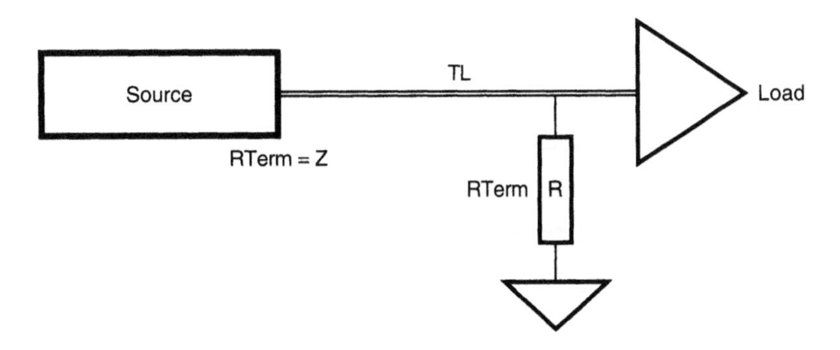

Figure 9.30

Parallel Termination

Thevenin termination is similar to parallel termination, but replaces the single termination resistor (Rterm) with two resistors, one for the positive load and one for the negative load. The resistors are connected to respective IOL and IOH sources.

Thevenin termination allows efficient absorption of the transient current, without the asymmetrical loading problem of the simple parallel termination. There is still the disadvantage of a constant DC load.

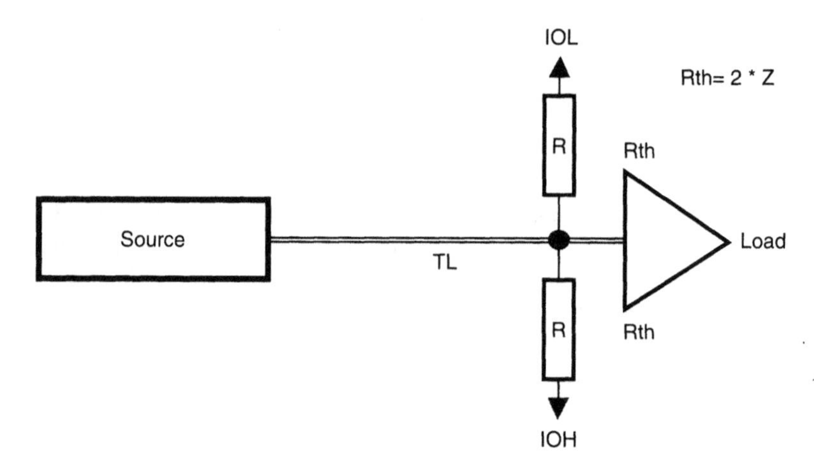

Figure 9.31

Thevenin Termination

Resistor–Capacitor termination is similar to parallel termination. A capacitor is placed between the termination resistor (Rterm) and ground to allow for a low impedance at AC without presenting a DC load. If all that a capacitor did was to block DC and pass AC, this would be an ideal solution. Unfortunately, another feature of a capacitor is that it will trap a charge. Because the capacitor stores a charge, it can distort the signal when discharging back into the transmission line.

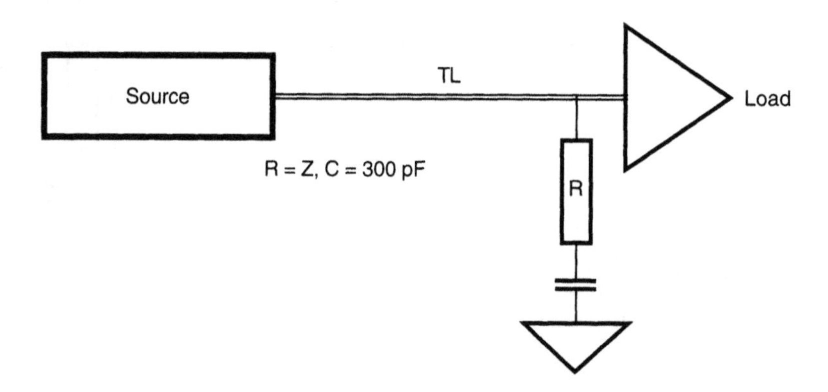

Figure 9.32

RC Termination

Diode termination does not attempt to match the impedance, but rather conducts the reflected voltage and current to ground or VDD. This is the most effective termination method when the load circuit is accessible. Diode termination has an additional advantage because the activation energy for a silicon diode has a voltage–current ratio of approximately 50 ohms. Is that cool, or what?

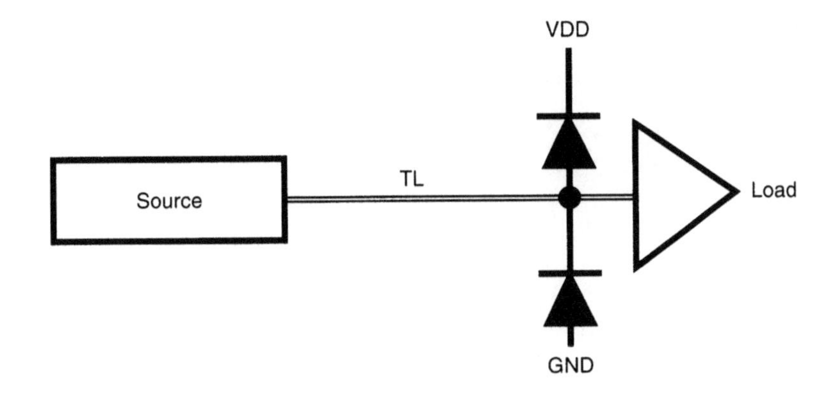

Figure 9.33

Diode Termination

Chapter Review Questions

1. Why is the test circuit design for a mixed signal device typically more demanding than for a digital logic device?

2. What is a ground loop?

3. What is a star ground?

4. What is the purpose of the power supply decoupling capacitors?

5. What is a transmission line? What are some effects of a mismatched transmission line?

Glossary

AC Tests Alternating Current tests pertain to verifying the dynamic switching characteristics of the device.

active load A programmable circuit that acts as a pull-up or pull-down load when connected to a DUT pin.

Active-Low A signal whose active state is considered to be a logic 0.

Active-High A signal whose active state is considered to be a logic 1.

ADC (Analog to Digital Converter) A device or process that converts analog input levels into digital values.

alias frequency A false lower-frequency signal component that is caused by an inadequate sample frequency.

anti-alias filter A filter that attenuates high-frequency components of an analog signal that are above one- half the sample frequency, prior to the analog-to-digital conversion.

aperture delay The delay between the command to start an ADC conversion and the time it actually occurs.

aperture jitter Operation variations in aperture delay.

ASCII (American Standard Code for Information Interchange) Represents text by binary numbers.

ATM (asynchronous transfer mode) A high-bandwidth networking standard which uses asynchronous time-division multiplexing.

attenuation Reduction of signal strength, measured in dB; negative gain.

autoranging The capability of an instrument to switch among ranges automatically.

AWG Arbitrary Waveform Generator.

band-pass filter A filter that passes only those signal frequencies between two set frequencies.

bandwidth (BW) A range of frequencies over which a device or instrument can properly process the signal.

bi-directional pins function as either inputs or outputs depending on the device mode.

bin A designated category that results from the completion of the device test.

binary search A test method that performs a successive approximation sequence by incrementing or decrementing progressively smaller binary-weighted values.

bipolar A semiconductor device that is manufactured with NPN and/or PNP transistors, as distinct from the Field Effect Transistors used in NMOS or CMOS processes.

BIST Built-in self-test.

clamps Programmed values that limit the amount of current or voltage than be supplied by a tester resource.

CMOS (Complimentary Metal Oxide Semiconductor) A semiconductor process by which a device is manufactured with both N-channel and P-Channel FET transistors.

CMRR (common-mode rejection ratio) A measure of a circuit's ability to reject signals common to both input connections, usually expressed in dB.

CODEC An abbreviation of coder-decoder; a device that can encode (convert from analog to digital) and decode (convert from digital to analog).

comparators generate a logic level based on the difference between the two input pins. Dual comparators are used by the pin receiver to detect logic levels from the device output pin.

conversion rate The rate at which a converter can generate valid output data or levels.

convolution The integration of the product of two functions in time. Convolution in the time domain is equivalent to multiplication in the frequency domain.

crest factor The ratio of the signal peak value to the signal RMS value.

crosstalk The unwanted transfer of a signal from one circuit to another

cycle time The time duration from the start of one cycle to the start of the next.

DAC (Digital to Analog Converter) A device or process that converts digital input values into analog levels.

datalog A report of the test results for a given device.

DDS (Direct Digital Synthesis) A signal-generation technique in which the signal is directly synthesized using only digital techniques.

decibel (dB) A logarithmic measure of the ratio of two signal levels: dB = 20 log (Voltage1 / Voltage2) and dB = 10 log (Power1 / Power2).

device LSB The expected step size for an ADC or DAC.

digital pertains to calculation or processing by numerical methods or by discrete units.

differential inputs Two inputs, where the measured signal is the difference between them.

distortion Periodic signal error, or an error in the signal shape. Distortion errors are multiples of the primary signal frequencies, also known as harmonics.

DNL (Differential Non-Linearity) On a DAC or ADC, the difference between the expected and actual step size.

DPS (Device Power Supply) The tester resource that provides power to the Device Under Test.

driver The section of the pin electronics that applies the VIL and VIH logic levels to the device input pins.

DSP A device or system that processes signal information in digital form.

DUT Device Under Test.

dynamic load Part of the receive section of the pin electronics that presents the IOL and IOH current loads to the device output pins.

dynamic range The ratio of a specified maximum level to the minimum detectable value; usually expressed in dB.

ESD Electrostatic discharge.

ethernet Popular LAN transmission network, based on a bus network topology, runs at 10 or 100Mb/s.

FET Field Effect Transistor.

FFT (Fast Fourier Transform) A mathematical method of analyzing the frequencies in a measured waveform.

filter A circuit or function that selectively removes noise from a signal. Electronic filters include low-pass, band-pass, and high-pass types. Mathematical filters can implemented with a set DSP algorithms.

format A description of the drive signal condition after the drive condition terminates. Common formats include NRZ (non-return to zero), RZ (return to zero), and CS (compliment surround).

gain The factor by which a signal is amplified, sometimes expressed in dB.

glitch A transient pulse in a circuit, narrower than the allowable functional pulse width.

ground An electrically neutral connection that has the same potential as the surrounding earth, used as the common return path and the zero-voltage reference point.

ground loop An unintentionally induced feedback loop caused by two or more circuits sharing a common electrical ground.

ground plane A large circuit board conductor or trace used as a common electrical reference point and ground return for circuits.

guard band (device testing) Adjustment made to a DUT's test specification to take into account test-system accuracy, repeatability, reproducibility, and correlation.

guard ring Consists of a guard conductor driven by a low-impedance source surrounding the lead of a high-impedance signal. The guard voltage is kept at the signal level to reduce current leakage effects.

high-impedance An output pin condition that effectively turns off the output logic level. The High-impedance mode is sometimes abbreviated as Hi-Z. A tri-state device pin can be at a logic low, a logic high, or high impedance.

high-pass filter A circuit or mathematical function that attenuates the low-frequency components of a signal, while passing the high frequency components

histogram A data structure that organizes information according to the number of events per value.

hysteresis A device input pin with different thresholds for a signal that is going from low to high (positive slope), and a signal that is going from high to low (negative slope).

IC Integrated circuit, also known as a 'chip.'

ICC (Current (I) Common Collector) The amount of current flowing into the device power pin, VCC, on a bipolar device.

IDD (Current (I) Device Drain) The amount of current flowing into the device power pin, VDD, on an NMOS or CMOS device.

IDDQ A test that checks for CMOS interconnect flaws by measuring the small amount of IDD current on a CMOS device power pin.

input pin A device connection point that receives logic level signals from an external source.

INL (Integral Non-Linearity) On a DAC or ADC, the deviation from the ideal transfer curve with zero gain and offset error.

I/O Input/output

IOH The amount of current that a device output pin must provide while maintaining an output level above the VOH (Voltage Output High) level.

IOL The amount of current that a device output pin must provide while maintaining an output level below the VOL (Voltage Output Low) level.

IOZ The maximum amount of current flow that can occur when a device pin is OFF (tri-state) in the high impedance mode.

IOZH The amount of IOZ current that flows with a logic high voltage level applied to a pin in the high impedance mode.

IOZL The amount of IOZ current that flows with a logic low voltage level applied to a pin in the high impedance mode.

ISDN (integrated services digital network) A telecommunication network standard for transmitting voice and data.

jitter Instability of a signal, effectively a form of frequency modulation or noise across time.

latch-up A uncontrolled high current condition in a CMOS device usually caused by reverse-biasing the substrate.

linear search An alternative to the binary search method. The linear search steps from the lower boundary to the upper boundary in consecutive increments. It is slower than the binary search, but has advantages for locating hysteresis boundaries.

logic The ability to generate a predictable output decision based on input states.

lot A quantity of devices which are grouped together for the purposes of processing and testing.

low-pass filter A circuit or function that attenuates the high-frequency components in an analog signal while passing the low-frequency components.

mask A method of disabling the compare function of the pin electronics receive circuit. A mask serves to ignore the device output logic state, which is called a "don't care" condition. Don't care is typically designated by an 'X' in the pattern file symbols.

mismatch A less than ideal coupling of two circuits that can create reflections and signal degradation.

modem An abbreviation of modulate-demodulate; a device that can transmit a signal by modulating a carrier signal; and receive a signal by demodulating a carrier signal.

modulation Transmitting a message by variations in amplitude, frequency, and/or phase of a carrier wave.

monotonicity A characteristic of a DAC in which the analog output increases as the digital input code increases.

MPEG (Motion Pictures Experts Group) Standard for digital compression of video and audio information.

multi-site testing Using a single test program and a single test head with more than one test site to test more than one device either in parallel or in sequence.

N-channel is a semiconductor material that has an excess amount of electrons. Applying a negative charge to N-channel material drives the electrons away and forms holes.

negative current flow The current direction convention for current flowing out of the device and into the test system.

NMOS (N Channel Metal Oxide Semiconductor) A semiconductor process by which a device is manufactured with exclusively N-channel transistors.

noise Undesired electrical signal information. For mixed signal test, noise is usually defined as random spurious signal information tha occurs at non-harmonic intervals of the signal frequency.

NRZ (non-return to zero) A digital pattern format that represents logic states simply by high and low signal levels, rather than pulses.

NTSC (National Television System Committee) US standard for analog color television broadcasting.

Nyquist A theorem that states that the sampling frequency must be greater than twice the bandwidth of the signal.

output pin A device output pin drives external circuits with logic voltage levels corresponding to the device's internal logic response.

output sample The point in time when the tester comparators are strobed to evaluate the logic output state of a device pin.

passive filter A filter circuit using only passive components such as resistors, capacitors, and inductors.

P-channel A semiconductor material that has an excess amount of holes. Applying a positive charge to P-channel material drives the hole away and attracts electrons.

per pin resources Resources that are dedicated to each pin, and do not require multiplexing or sharing.

Pin Electronics, PE, PEC Pin Electronics is the circuitry that interface the device logic pins to the tester resources. The PE circuits include drivers, comparators, and programmable load circuits, as well as a means of connecting to DC and AC instrumentation.

PLL (Phase Locked Loop)

PMU (Precision Measurement Unit) A DC measurement system that can force and measure voltage or current with a high level of accuracy.

positive current flow The current direction convention for current flowing out of the test system and into the device.

POTS Plain Old Telephone Service.

power pin A device pin that is connected to power or ground.

propagation delay The period of time between a valid device input condition and a valid device output condition.

quantizing error The inherent uncertainty in representing analog levels with discrete digital values, caused by the finite resolution of the conversion process. Increasing the resolution of an ADC reduces the quantizing error.

quantizing noise The effects of quantizing error when digitizing a dynamic signal.

range The maximum and minimum allowable full-scale signal (input or output) that yields a specified performance level.

receiver The section of the pin electronics that detects the logic levels from the device output pins.

resolution The smallest division to which a measurement can be determined. For example, an ADC with 12-bit resolution can resolve to 1 part in 4096.

rise time Period of time required for the positive slope of a pulse to change from 10% to 90% of its final value.

RMS (root mean square) A form of averaging calculated by dividing the sum of squares by the number of samples, and then taking the square root.

RZ (return to zero) A digital pattern format in which binary 1 symbols are represented by a pulse rather than a level.

Sample and Hold Amplifier (SHA) A circuit that acquires and temporarily stores an analog level (usually with a capacitor).

SEMI Semiconductor Equipment and Materials International Association.

settling time The period required after a signal transition for signal level to settle and remain within a specified error band around the final value.

SFDR (spurious-free dynamic range) The span between the signal amplitude and the amplitude of the worst-case error; usually expressed in dB.

shared resources are centralized resources that can be connected to device pins through a relay matrix.

SINAD (Signal to Noise and Distortion Ratio) The ratio between the signal amplitude and the total amplitude of the noise error and the distortion error, usually expressed as a dB.

sink Current that flows from the test system into a device pin. Sink current is the same as negative current.

slew rate Rate of change for signal transition, calculated as the change in voltage divided by the change in time.

Shmoo plot A graph representing test results across variations of two variables.

SNR (Signal to Noise Ratio) The ratio between the signal amplitude and the total amplitude of the noise error, usually expressed as a dB. The conventional measurement method for SNR does not include the distortion error.

source Current that flows from a device pin into the test system. Source current is the same as positive current.

spectrum analyzer An instrument that displays the frequency spectrum of a signal.

strobe The event that acquires the DUT output pin logic state.

successive approximation see Binary Search

summary is a report of the test results for a group of devices, such as a lot or a wafer.

tester-per-pin A test system designed with a full set of per-pin resources.

Total Harmonic Distortion (THD) The algebraic sum of all of the distortion error across a given bandwidth.

tri-state A device pin that functions either as an input, and output, or a high impedance (off) connection.

UUT (Unit Under Test) The system test equivalent of a Device Under Test.

VCC (Voltage Common Collector) The supply voltage for a bipolar device.

VCO A voltage-controlled oscillator.

VDD (Voltage Device Drain) The supply voltage for a NMOS or CMOS device.

vector A line in the pattern that corresponds to specific input and output logic conditions for a given test cycle.

VREF The reference voltage for the dynamic load circuit. The dynamic load will apply the IOH current load to a device output level above VREF, and will apply the IOL to a device output level below VREF.

VSS (Voltage Source Supply) The device ground pin.

wafer A silicon disk that has been processed to contain integrated circuits. The individual circuits on a wafer are called a die.

wafer map A graphic display of the test results for each die on wafer, plotted according to the die location.

X A pattern notation for a 'don't care' condition.

Z A pattern notation for a high-impedance (off) condition.

ZIF socket A zero-insertion-force DUT socket

INDEX

P

Q

R

S

T

CPSIA information can be obtained
at www.ICGtesting.com
Printed in the USA
JSHW020954101219
2893JS00002B/115